Front cover: The Monitor. From *Hydraulic Mining in California*, by Taliesin Evans. The Century Magazine volume 25, number 3, January 1883.

GEOLOGY AND MINERAL DEPOSITS OF THE COLFAX AND FORESTHILL QUADRANGLES, CALIFORNIA

By DEB K. CHANDRA
Geologist, India Geological Survey

SPECIAL REPORT 67
CALIFORNIA DIVISION OF MINES
FERRY BUILDING, SAN FRANCISCO, 1961

SEAL OF THE

EUREKA

STATE OF CALIFORNIA
EDMUND G. BROWN, Governor

DEPARTMENT OF NATURAL RESOURCES
DeWITT NELSON, Director

DIVISION OF MINES
IAN CAMPBELL, Chief

Special Report 57
Price 9 P.50

CONTENTS

Illustrations

Tables

LETTER OF TRANSMITTAL

EDMUND G. BROWN
Governor of the State of California
Sacramento, California

SIR: I have the honor to transmit herewith Special Report 67, *Geology and Mineral Deposits of the Colfax and Foresthill Quadrangles, California,* prepared under the direction of Olaf P. Jenkins, then Chief (now retired) of the Division of Mines, Department of Natural Resources. The Report is accompanied by detailed colored geologic and economic mineral maps, charts, and cross sections, as well as numerous diagrams and photos. The area covered in the report covers 130 square miles of the central Sierra Nevada, centering about 35 airline miles west of Lake Tahoe. Within its boundaries are the famous gold placer-mining districts of Iowa Hill, Foresthill Divide, Michigan Bluff, and many others, as well as numerous lode-gold mines. Other mineral deposits include chromite, asbestos, soapstone, and limestone.

The author, Dr. Deb K. Chandra, staff geologist in the Oil and Gas Commission of the India Geological Survey, did geologic work on the Colfax-Foresthill area as a partial requirement toward the degree of Doctor of Philosophy conferred upon him by the University of California at Berkeley in 1953. In addition to this doctorate, the author holds Bachelor of Science and Master of Science degrees from the University of Calcutta. Dr. Chandra's report is an important contribution to the all too meager geologic literature on this part of the Sierra Nevada.

Respectfully submitted,

IAN CAMPBELL
State Mineralogist and
Chief of the Division of Mines

PREFACE

Dr. Chandra's work in the Colfax and Foresthill quadrangles is the first detailed mapping published for this area since the Colfax folio by Waldemar Lindgren became available in 1900. Moreover, it marks the first attempt to decipher the complicated structural features of the area with the aid of petrofabric procedures and hence is a notable contribution to the literature on the structural geology of the central Sierra Nevada of California.

Unfortunately, the Division of Mines lacked the funds to process the Colfax-Foresthill paper until long after Dr. Chandra had returned to his native India. Consequently, we have been unable to bring all of the information on mineral deposits up to date. We have also been unable to make the stratigraphic presentations as complete and as lucid as we would like to have had them. Nonetheless, we believe our readers will find a great deal of useful and interesting material in the following report.

OLIVER E. BOWEN, JR.
Geologist, Division of Mines
San Francisco, March 1960

ABSTRACT

Approximately half of the Colfax-Foresthill area is underlain by rocks of the Upper Paleozoic Calaveras group; nearly one-fourth by stratified rocks of Jurassic age; and roughly one-fourth by mid-Tertiary fragmental volcanic rocks. Lesser patches of intrusive granitic and serpentinic rocks of Jura-Cretaceous age are present as well as small erosional remnants of Eocene auriferous gravel. A wide variety of minor hypabyssal and plutonic intrusive masses are broadly distributed over the map area.

The Upper Paleozoic Calaveras group is here divided into four formations which are, bottom to top, the Blue Canyon, Tightner, Cape Horn, and Clipper Gap. Shale, sandstone, and graywacke make up the bulk of the Blue Canyon formation; basic lava flows, tuffs, and agglomerates with minor chert and shale form most of the Tightner formation; shale, sandstone, and basic volcanic rocks predominate in the Cape Horn; and shale, sandstone, chert, conglomerate, and minor limestone make up the bulk of the Clipper Gap formation. The total thickness of the Calaveras group is 37,400 feet. Both the Tightner and Cape Horn formations exceed 14,000 feet in thickness.

The Logtown Ridge formation, lower Upper Jurassic (?), consists of about 2,000 feet of basic flows, tuffs, and agglomerates. Neither the top nor bottom is exposed.

The Mariposa formation, of late Jurassic age, consists of 14,600 feet of breccia, conglomerate, shale, siltstone, sandstone, graywacke, and grit with penecontemporaneous basic intrusions. It lies unconformably on the Clipper Gap formation over part of the map area.

Diabase and albite diabase are common intrusive rock types in the Clipper Gap and Mariposa formations. Serpentinized peridotite and gabbro cut a wide variety of older rocks. A single, oval mass of granodiorite in the northwest corner of the map area is the only felsic granitic intrusion present of notable size.

Lying unconformably on the upturned edges of the complexly deformed basement rocks are three relatively undeformed rock units of Tertiary age. Lowermost are the auriferous gravels of middle Eocene age, consisting, in most places, of bottom or deep channel gravels and bench gravels. The formation does not exceed 150 feet in thickness anywhere in the map area. Vein quartz and metachert debris predominate over other bedrock types in the auriferous gravels. Stratigraphically above the auriferous gravels is the Valley Springs formation of probable late Miocene age, which reaches a maximum thickness of 400 feet. This formation is succeeded unconformably by the Mio-Pliocene Mehrten formation consisting of andesite tuffs, breccias, sands, gravels, and boulder beds reaching 700 feet in thickness.

Rocks of the Calaveras group were subjected to two deformations, the first prior to deposition of the Jurassic formations and the second during the Nevadan orogeny. The Calaveras rocks which have suffered two orogenies have been deformed into overturned, isoclinal folds trending north-northwest, whereas the Jurassic formations which have suffered a single orogeny are rarely overturned. Several strike faults, mostly of thrust type, cut the map area. These commonly dip east at angles of between 60 and 80 degrees.

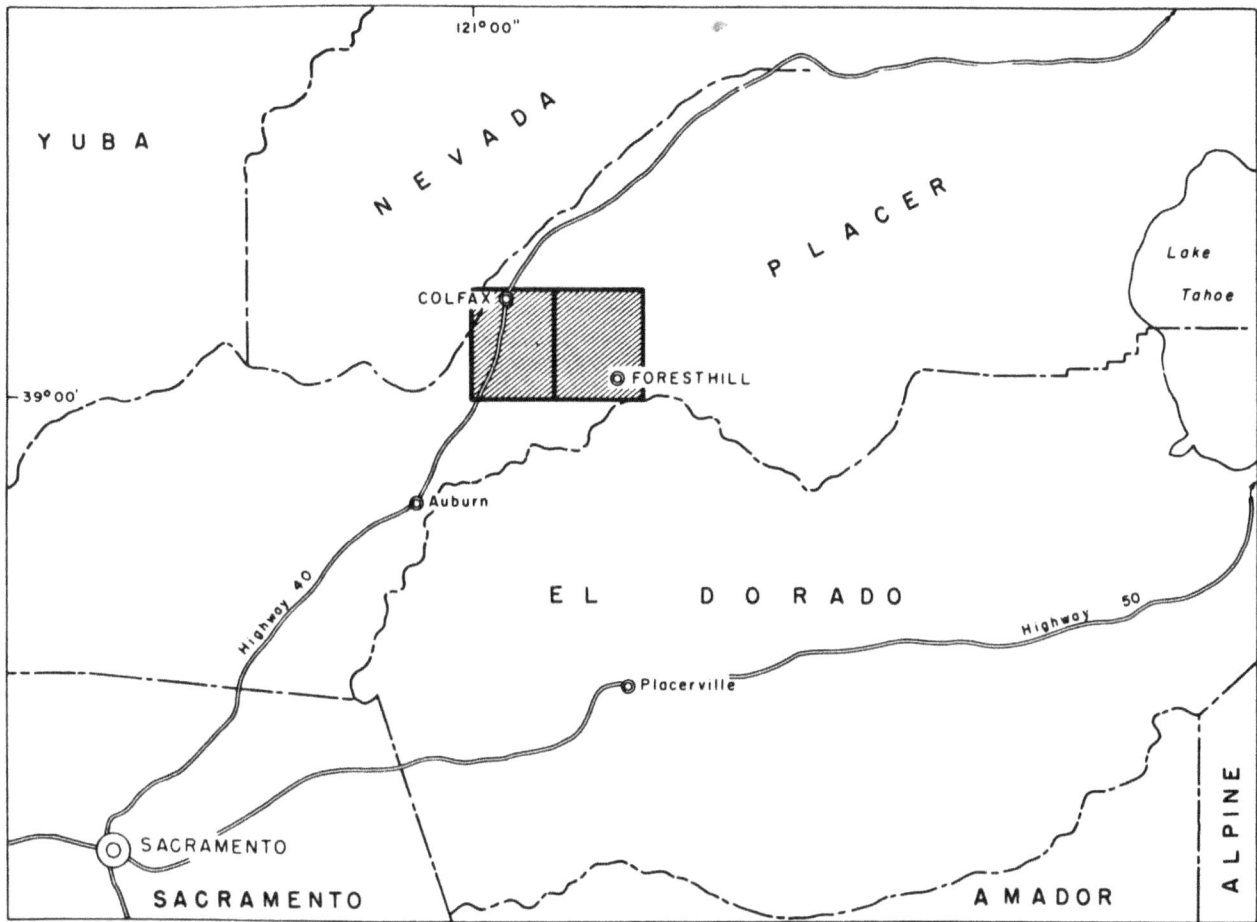

FIGURE 1. Index map showing the location of the Colfax and Foresthill quadrangles.

GEOLOGY AND MINERAL DEPOSITS OF THE COLFAX AND FORESTHILL QUADRANGLES, CALIFORNIA

By Deb K. Chandra

INTRODUCTION

Location and Access.—The area studied is bounded by the parallels 39° 00' and 39° 07' 30" north, and the meridians 120° 45' and 121° 00' west. It lies principally in Placer County, California; and extends northwestward and southeastward into Nevada and El Dorado Counties respectively. Mapping was confined to two adjoining 7.5-minute quadrangles, Colfax and Foresthill. The area is traversed by Highway U. S. 40.

Field Work. Field work was done during the fall of 1950, the summer of 1951 and parts of the summer and fall of 1952. Relevant topographic maps on the scale of 1: 24,000 were under preparation when the work was started. Mapping was done on aerial photographs taken in 1938 for the U. S. Department of Agriculture, Division of Forestry. The topographic maps (edition of 1951) which form the base for the geologic map of this report were drawn from aerial photographs taken in 1946.

Scope of Thesis. The objectives of this investigation are: to map the geology of the area, to describe the stratigraphy and petrography of the different formations, and to interpret the geologic structure.

There are numerous mines in the area. Most of them were abandoned long before World War II, but others were closed during the war. Many have caved in or are covered by debris or landslides. Detailed information about them is largely lacking and most of the general information about them listed here has been gleaned from the literature.

Scope of Laboratory Work. Composition of plagioclase feldspars when stated specifically (for example, An$_{3-4}$) has been determined by standard Universal Stage method. Determination of the optic axial angle and extinction angle of pyroxenes has been done in twinned crystals on a Universal Stage. The extinction angles of amphiboles have been found from sections normal to Y.

Previous Work. The area under review forms the southern half of the southwest quadrant of the Colfax quadrangle (scale 1: 125,000), the geology of which was mapped by Lindgren and published in Folio No. 66 of the U. S. Geological Survey (1900). Later Lindgren (1911) described the auriferous gravels and reconstructed the Tertiary channels. In 1917 Moody published a detailed study of the breccias of the Mariposa formation. Some of Knopf's (1929) general statements about the Mother Lode district are also applicable to this area. Tectonic patterns in the Sierran belt have been discussed by

Mayo (1937 and 1941), and Locke, Billingsley, and Mayo (1940). Studies on fossils in Tertiary beds in the area and its neighborhood have been done by Knowlton (Lindgren, 1911, pp. 54-64), Chaney (1932), Potbury (1937), MacGinitie (1941), and Condit (1944). Reports on the mineral deposits include State Mineralogist's Reports (1890, 1936, 1943), and those by Logan (1934, 1936), Averill (1946), and Rynearson (1953).

Acknowledgements. The writer is indebted to Professors Francis J. Turner and Garniss H. Curtis for advice and guidance, and to Professor Howel Williams for valuable discussions. Thanks are also due to Mr. M. D. Turner of the State Division of Mines, California, for suggesting some areas for investigation, and to Mr. Oliver Bowen of the same source for valuable discussion in the field. The writer finally wishes to thank the Department of Geological Sciences of the University of California for financial assistance in the course of field work.

GEOGRAPHY

The Colfax and Foresthill quadrangles lie between the western foothills and the crest of the Sierra Nevada at an average elevation of 2700 feet. A profile from west to east shows a gently rolling surface, in many places an even tableland, with a slight westerly slope in which streams have cut deep V-shaped canyons to depths as great as 2000 feet. The major part of the area is drained by the North Fork of the American River, and minor areas in the northwest and southeast are drained by the Bear River and Middle Fork of the American River respectively.

STRATIGRAPHY

The oldest rocks within the area belong to the Calaveras group of late Paleozoic age. There is a big stratigraphic break between the end of the Paleozoic era and the upper Jurassic epoch. The first record after the Calaveras group lies in the Logtown Ridge formation. Deposition of the Mariposa formation followed. These Paleozoic and early Mesozoic formations were intruded by basic rocks before the onset of Nevadan orogeny. Folding and faulting accompanied the emplacement of ultra-basic intrusive rocks, gabbro, and granodiorite. Intrusion of lamprophyre and keratophyre and development of quartz veins also took place during this period. All the events, beginning with deposition of the Logtown Ridge formation and ending with the emplacement of plutonic and intrusive bodies, took place during the Upper Jurassic epoch. The next stratigraphic record lies in the "auriferous gravels"—equivalent to the Ione formation of middle Eocene age. The Valley Springs formation is next youngest followed by the uppermost Mehrten formation.

Calaveras Group

Lindgren (1960) subdivided the Calaveras group into five formations in the larger Colfax quadrangle; of these, the following three, numbered in order of decreasing age, are represented in the area under review: Clipper Gap formation, Cape Horn formation, and Blue Canyon formation. The writer has differentiated another formation represented by meta-andesite, amphibolite, hornblende-schist, and actinolite-schist. The stratigraphic position of this formation is between the Blue Canyon and Cape Horn formations. The lithology of this formation closely resembles that of the Tightner formation of Ferguson (1929) in the Alleghany district in the northern part of the larger Colfax quadrangle; and the stratigraphic position is the same. This name will therefore be used here. The downward sequence of formations recognized in this paper are as follows: Clipper Gap formation, Cape Horn formation, Tightner formation, Blue Canyon formation.

Blue Canyon Formation

The Blue Canyon formation is exposed in the southeast and northeast corners of the map area. It consists principally of interstratified slate, sandstone, graywacke, grit, and conglomerate. Fissile slate is predominant and is locally rich in pyrite. Sandstone and graywacke are massive and form steep cliffs. The bands of grit and conglomerate are few and nonpersistent. Quartz-mica schist, though rarely found, is confined to the Blue Canyon formation. Several thin sills of andesite are intrusive into this formation. These intrusions are probably contemporaneous with the volcanic rocks of the Tightner formation. The formation reaches a thickness of 5100 feet.

Graywacke is massive and nonstratified, and shows lack of sorting. The rock is composed of large angular detrital grains—chiefly quartz, microcline, orthoclase, plagioclase (mainly oligoclase), muscovite, biotite, and iron ore. The groundmass is made up of sericite and chlorite which may have been formed from the clay matrix by low-grade metamorphism.

Sandstone is thick-bedded, medium- to coarse-grained rock. It shows poor sorting. The grains are subrounded or angular, but the larger grains are often rounded. The rock consists of quartz, orthoclase, plagioclase (mainly oligoclase) and muscovite in a groundmass of quartz, sericite, and carbonate minerals.

Lindgren (1900, p. 2) reported two fossiliferous limestone lenses in the Blue Canyon formation several miles east of the present area. The fossils are poorly preserved,

and are: *Lithostrotion, Aviculopecten, Murchisonia, Syringopora, Diphyphyllum,* and crinoid stems. In his opinion, they indicate nothing more definite than a Paleozoic age.

Tightner Formation *

The Tightner formation occupies a major part of the eastern half of the area, and is more than 14,000 feet thick. It is composed principally of volcanic materials—flows, tuffs, and agglomerates with occasional intercalations of argillaceous and siliceous sediments. The chief rock types are hornblende-schist, amphibolite, and meta-andesite; slate, chert, and dacitic tuff form minor constituents, and occur as thin nonpersistent intercalations.

The relationship between the main body of the Tightner formation and the Blue Canyon formation is obscured by the intervening presence of large serpentine bodies. Meta-andesite occurs as sills in the Blue Canyon formation and resembles closely the meta-andesite of the main body of the Tightner formation. The amphibolite rocks exposed as inliers in the Clipper Gap formation are also presumed to belong to the Tightner formation on the basis of lithologic similarity.

Hornblende-schist is the most common rock type in the formation. The predominant mineral is hornblende partially replaced by actinolite or chlorite along the crystal edges. Plagioclase is invariably saussuritized. Other minerals constituting the rock are epidote, clinozoisite, iron ore, titanite, and leucoxene. Locally, the hornblende is completely replaced by actinolite and chlorite, indicating a retrograde metamorphism.

Amphibolite has a somewhat granulitic texture, distinct from that of the hornblende-schist just described. It consists of hornblende, albite, clinozoisite, iron ore, titanite, rutile, and occasionally quartz, epidote, calcite, prehnite, and often sulphides. The prehnite is present as thin veinlets in the rock. Some of the amphibolite specimens contain garnet. The granulitic texture and the occasional presence of garnet indicate a grade of metamorphism higher than is normal for amphibolite, but occasionally these rocks are brecciated and show subsequent alteration. In these, hornblende is partially or wholly replaced by actinolite, garnet by chlorite, and in one case actinolite by stilpnomelane. These alterations indicate a later retrograde metamorphism to which the amphibolite was subjected under the shearing stress prevailing at the time of brecciation.

Meta-andesite is less common in the Tightner formation than either hornblende schist or amphibolite. It consists of hornblende, albite, epidote, clinozoisite, ilmenite, titanite, sulphides, and chlorite. Ilmenite is partially replaced by leucoxene and titanite. The rock has an in-

---

* The author in his original Ph.D. dissertation introduced the name Shirttail Canyon for this formation but inasmuch as the supporting data were incomplete the approximately equivalent formational name Tightner, already in use in the Colfax vicinity, was substituted by the editors.

cipient schistosity with the hornblende crystals showing some preferred orientation.

Meta-dacite is fine- to medium-grained, and consists of quartz, microcline, albite, hornblende (partially replaced by actinolite), biotite, titanite, chlorite, epidote, clinozoisite, and limonite. The minerals are distributed in alternate mafic and felsic bands.

Meta-chert contains banded and spherulitic chalcedony, sericite, and limonite. Veinlets of quartz are also present.

Graphite-schist (occurring as thin inclusions in meta-andesite near Volcano Canyon) consists of quartz, biotite, albite, orthoclase, titanite, rutile, leucoxene, sericite, and abundant flaky graphite.

Cape Horn Formation

The Cape Horn formation covers about one-third of the area and reaches a thickness of 14,900 feet. It is composed predominantly of fissile slate with several bands of sandstone, graywacke, chert, grit, and conglomerate. These sediments are interbedded with volcanic materials which occur as flows, tuffs, and agglomerates. The intense volcanism which characterized the period represented by the Tightner formation had partially subsided when the Cape Horn formation began to be deposited. During the later part of deposition of the Cape Horn formation, however, volcanism again became moderately intense. Locally the volcanic ashes are so intermixed with argillaceous sediments that they cannot be mapped separately. The volcanic ashes where not too severely metamorphosed may show faint graded bedding and local microfolding and microfaulting. The volcanic rocks are represented by meta-basalt, actinolite-schist, and rarely tremolite-schist.

Slates of the Cape Horn formation are similar in appearance and mineralogy to other slates in the area, having as their principal components—quartz, chlorite, sericite, biotite, iron ore, limonite, titanite, and some fine clay minerals. One thin section shows extensive development of porphyroblasts of rutile.

Sandstones locally show graded bedding. They are often silicified.

Grits and conglomerates have a slaty matrix and show an incipient schistosity with the pebbles slightly flattened in the plane of schistosity.

Meta-basalt is fine- to medium-grained and occasionally porphyritic with phenocrysts of augite and plagioclase. Fine-grained meta-basalt consists of albite (An_{2-5}), hornblende, epidote, clinozoisite, actinolite, titanite, iron ore, and quartz. Hornblende is replaced by actinolite. Augite, if present, is invariably uralitised. Two thin sections contain stilpnomelane developed at the expense of actinolite. The meta-basalt is often cut by thin veinlets of quartz, carbonate or chlorite-titanite. In most of the thin sections the hornblende prisms show a preferred orientation and the rocks have an incipient schistosity. The porphyritic meta-basalt is chiefly composed of phenocrysts of augite ($2V = 55°$; $Z \wedge c = 43°$) and plagioclase. Augite is partially or completely uralitised and plagioclase is saussuritized. The groundmass consists of hornblende, titanite, limonite, epidote, clinozoisite, albite, and carbonate. Amygdules, if present, are composed of chlorite and calcite.

Actinolite-schist is markedly schistose and consists of actinolite, chlorite, albite, clinozoisite, titanite, and iron ore. Actinolite is partially replaced by chlorite.

Tremolite-schist is fibrous and is composed of tremolite, chlorite, talc, carbonate, clinozoisite, titanite, and albite.

Five samples of metavolcanic rocks belonging to the Cape Horn formation were analyzed. Of these, two were fine-grained, two medium-grained, and the fifth coarse-grained. The analyses (Nos. 1 to 5 in Table 1) show some common chemical features such as low silica, high alumina, low titania and iron-oxide, high lime, and almost equal amounts of soda and potash. They show a close similarity with the analyses of Nos. 6 to 9 in Table 1, except in their alkali content; the latter have higher soda and lower potash contents. No. 6 and probably No. 9 are of the same age as the metavolcanic rocks under discussion; and Nos. 7 and 8 are probably contemporaneous or slightly younger but petrographically similar. No. 10, an average basalt, shows close similarity with Nos. 1 to 5, except in its higher soda content. When compared with No. 11, an average andesite, Nos. 1 to 5 show much lower silica but higher magnesia and lime. On the whole, rocks 1 to 5 chemically resemble more closely No. 10 (average basalt) than No. 11 (average andesite). According to current general opinion based on megascopic and microscopic appearance, volcanic rocks similar to rocks 1 to 5 have been called andesite. As is evident from the chemical analyses, these rocks are basaltic; their high alumina and potash content is somewhat unusual.

Variation diagrams (fig. 2) for different oxides plotted against silica, are drawn in the case of the five basic metavolcanic rocks. Smooth curves are generally obtained, though there are a few stray points lying close to the curves.

Lindgren (1900, p. 2) reported one "little limestone mass" containing "round crinoid stems, probably indicating a Paleozoic age." No such mass nor any other fossiliferous material has been found by the writer. As the Cape Horn formation is bounded on both sides by faults no more definite estimate of its age can be given.

Clipper Gap Formation

The Clipper Gap formation occupies the western part of the area and is more than 3000 feet in thickness. Besides the main body there are several other narrow belts of the formation bounded by fault planes. The formation

Table 1. Chemical analyses of Paleozoic volcanic rocks from the Colfax quadrangle, compared with analyses of rocks from adjacent areas and averages.

	1.	2.	3.	4.	5.	6.	7.	8.	9.	10.	11.
SiO₂	49.48	48.36	49.30	48.86	46.35	49.10	50.00	48.86	48.00	49.06	59.59
Al₂O₃	20.66	16.82	15.96	18.12	16.07	15.50	13.40	--	15.00	15.70	17.31
FeO	6.27	7.68	7.84	6.97	8.34	8.40	11.70	--	9.00	6.37	3.13
Fe₂O₃	1.16	1.74	1.43	2.83	2.70	--	--	--	5.00	5.38	3.33
TiO₂	0.48	0.80	0.62	0.81	0.68	--	0.06	--	--	1.36	0.77
MnO	0.17	0.34	0.21	0.26	0.19	--	--	--	--	0.31	0.18
CaO	6.44	10.27	9.06	9.02	12.03	8.50	11.00	7.65	9.00	8.95	5.80
MgO	5.89	5.48	8.44	4.88	6.77	11.00	7.00	9.88	6.00	6.17	2.75
K₂O	2.33	2.83	1.96	2.09	1.68	1.50	0.40	1.36	0.50	1.52	2.04
Na₂O	2.92	1.54	2.15	2.44	1.22	3.60	3.10	3.46	4.00	3.11	3.58
H₂O−105°C	0.38	0.16	0.18	0.48	0.27	1.90	2.20	--	--	1.62	1.26
H₂O+105°C	3.64	3.07	2.72	2.97	3.04	--	--	--	--	--	--
CO₂	0.00	0.35	0.00	0.00	0.00	--	--	--	--	--	--
P₂O₅	0.10	0.39	0.08	0.33	0.45	--	--	--	--	0.45	.026
Total	99.92	99.83	99.95	100.06	99.79	99.50	99.86	--	--	100.00	100.00

Analyst: W. H. Herdsman.

1. Fine metavolcanic tuff, east bank, North Fork, American River. In Sec. 12, T 14 N., R. 9 E . . . Cape Horn formation.
2. Fine metavolcanic tuff, east bank, North Fork, American River, NW ¼ of Sec. 19, T. 14 N., R. 10 E . . . Cape Horn formation.
3. Medium-grained metavolcanic flow, east bank, North Fork, American River. SW ¼ of Sec. 1, T. 14 N., R. 9 E . . . Cape Horn formation.
4. Medium-grained metavolcanic flow, south bank of Indian Creek near its mouth. SW corner of Sec. 18, T. 14 N., R. 10 E . . . Cape Horn formation.
5. Porphyritic coarse-grained metavolcanic flow, east bank, North Fork, American River. NW ¼ of Sec. 19, T. 14 N., R. 10 E . . . Cape Horn formation.

6. Augite crystal tuff in Melones slate (Calaveras formation), Carson Hill, Calaveras County, California. (Moss, Frank A., 1927).
7. Uralite porphyry in Melones slate. (Age of uralite porphyry unknown). (Moss, Frank A., 1927).
8. "Augite porphyrite containing squarish hornblende paramorphic after augite" from Plumas County. (Age unknown.) (U.S. Geol. Survey, 14th An. Rep., Pt. 2, p. 473, 1892-93.)
9. Plagioclase amphibolite (approximate chemical composition, calculated from mineralogical analysis), Fancher Creek, Dinuba quadrangle, California. (Macdonald, G. A., 1941, p. 231.) (The rock is probably of Calaveras age.)
10. Average basalt (average of 198 analyses, including 161 basalts). (Daly, R. A., 1933, p. 17.)
11. Average andesite (average of 87 andesites). (Daly, R. A., 1933, p. 16.)

consists of shale, argillaceous sandstone, grit, conglomerate, chert, occasional limestone, and metamorphosed shale. No contemporaneous volcanic rocks have been noted. Evidently the submarine volcanism which prevailed after deposition of the Cape Horn formation, had died out before the Clipper Gap formation was deposited.

Conglomerate is a conspicuous constituent of the Clipper Gap formation; it is widely distributed in thick but rather nonpersistent bands. Thin-bedded cherts with intercalations of slate are extensive. The thickness of the chert beds varies from one inch to ten feet or more, but beds are usually four or five inches thick. Most of the wide chert bands shown on the geologic map are made up of several bands so close together that they can hardly be shown separately on a map on the scale of 1: 24,000. A few bands of chert interbedded with manganiferous slates have been observed along the east-bound railway track about half a mile south of the Ponderosa road crossing. Two lenses of limestone belonging to the Clipper Gap formation occur along the eastern bank of the Bear River. The larger body of limestone is intruded by diabase of late Jurassic age.

Limestone is massive and well jointed. It is bluish gray and shows a saccharoidal texture in thin section. It has been reported to be fossiliferous but no fossils were found by the writer.

Conglomerate consists chiefly of ellipsoidal pebbles of chert in a slaty matrix, but pebbles of shale and sandstone are also present. A thin section of conglomerate shows rounded pebbles of chert, quartz, quartzite, sandstone, and andesite in clay matrix. (Presence of the chert pebbles suggests: existence of chert in a formation older than the Calaveras group, unconformity between the Clipper Gap formation and the older members of the Calaveras group, or subsequent silicification of the original pebbles in the conglomerate.)

The fossiliferous limestone, already mentioned, was reported by Lindgren (1900) to contain "crinoid stems, *Clisiophyllum gabbi* Meek, *Lithostrotion whitneyi* and fragments of various species of brachiopods." In his opinion, "these are the best Carboniferous fossils thus far obtained from this region and they can unhesitatingly be referred to the Lower Carboniferous."

Logtown Ridge Formation

The Logtown Ridge formation (Taliaferro, 1943) covers about one square mile of area near the south border of the map area on the eastern side of the North Fork of the American River. It is bounded on the east, north, and west by thrust faults dipping about 70° to the east. The formation consists of basic flows, tuffs, and agglomerates, and is at least 2000 feet thick. It forms an asymmetric anticline, steeply dipping on the east side. Flows predominate in the formation. Massive agglomerate beds are exposed at the Big Bend of the North Fork of the American River. All the rocks are less altered than similar rocks of the Calaveras group. Minor clay-slate beds intercalated in the volcanic formation are in the same stage of metamorphism as the Mariposa slate,

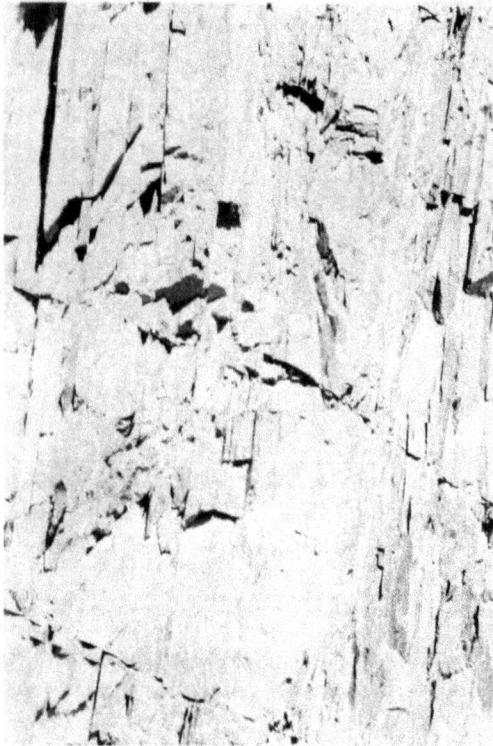

Photo 1 (Left). Steeply dipping thin-bedded sandstone of the Blue Canyon formation, Mosquito Ridge road.

Photo 2 (Above). Cape Horn formation: fine-bedded sandstone intercalated with slate. From North Fork of the American River. The 5-inch leather case is for scale.

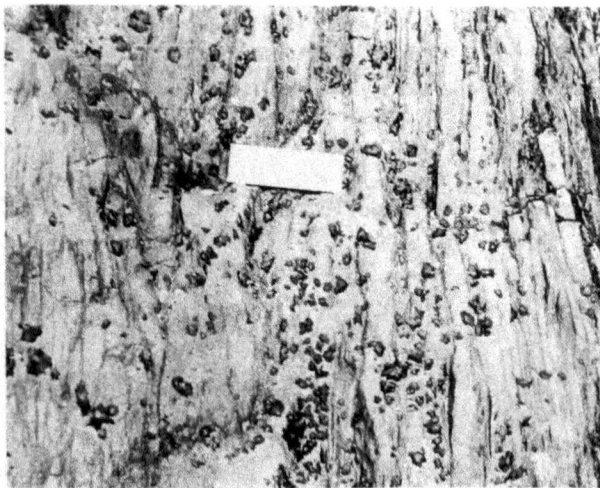

Photo 3. Thin-bedded chert of the Cape Horn formation, from Indian Creek near Iowa Hill road crossing. The 6-inch ruler is for scale.

Photo 4. Highly jointed meta-basalt of the Cape Horn formation from North Fork of the American River.

Photos 5 (above) and 6 (right).   Fine volcanic ash showing graded bedding and small-scale folding and faulting, Cape Horn formation, from North Fork of the American River. The 6-inch ruler is for scale.

Photos 7 (above) and 8 (right). Volcanic agglomerate of the Cape Horn formation, from North Fork of the American River. Baseline of the photos is about 6 feet.

but have not developed the typical phyllitic appearance of the Calaveras slates.

The volcanic rock is massive and fine- to medium-grained. It is generally porphyritic with phenocrysts of augite ($2V = 54°$; $Z \wedge c = 42°$) and plagioclase in a groundmass of plagioclase, actinolite, chlorite, titanite, and leucoxene. Phenocrysts of augite show zoning and hourglass structure, and are sometimes uralitised along borders, or are partially replaced by chlorite and/or calcite. Phenocrysts of plagioclase are usually saussuritized. Amygdules are abundantly present and are lined with chlorite. In all the thin sections the texture is well preserved, though most of the minerals are hydrothermally altered. Two of the thin sections collected in the Logtown Ridge formation are of volcanic breccia containing fragments of glassy basalt.

An average sample of fine-grained rock from a flow about 15 feet thick interbedded with medium-grained basic flows was analyzed (Analysis No. 1 in Table 2).

Rocks No. 1, 2, 4, and 5 show common chemical features. A thin section of rock No. 1 shows a basaltic texture—the rock is composed chiefly of albite and augite which are free from any alteration. It resembles a spilite. The high content of lime may have resulted from the presence of thin carbonate veins in the rock.

The medium-grained porphyritic rock which represents most of the flows of the Logtown Ridge formation

is andesite, according to normal classification procedures based on megascopic and microscopic appearance. The fine-grained rock (No. 1 in Table 2) constitutes a single flow interbedded with flows of porphyritic rock. Chemical analyses of rocks belonging to the Logtown Ridge formation elsewhere are not available. It is not known if any analysis has been done. Analysis No. 6 probably represents a rock from the Logtown Ridge formation and is a typical diabase or basalt; the results agree well with those of No. 1 (which is a typical spilite) except in soda content.

An analysis of medium-grained porphyritic rocks of the Logtown Ridge formation in the present area would be necessary in order to ascertain the chemical nature of the rock constituting the bulk of the formation.

From the relatively weak metamorphism exhibited by clay-slate interbedded with the flows, and from the well-preserved texture of the rock constituting the formation, the volcanic formation is considered to be younger than the Calaveras formation. On the other hand, it is stratigraphically below the immediately adjacent Mariposa formation. Taliaferro (1943, p. 284) reports several distorted ammonites in the Logtown Ridge formation elsewhere; but these fossils indicate nothing more definite than a Jurassic age. The fact that no *Aucellas* are associated with ammonites in the formation would indicate that the Logtown Ridge formation is older than Oxfordian, and is probably lower Upper Jurassic.

Mariposa Formation

The Mariposa formation covers a large part of the area. Besides the main body, there are several thin belts of this formation in the west and northwest. This repetition has been brought about by faulting and thrusting. The beds have been strongly folded, but the folds are rather open and are rarely overturned except in the south.

The formation consists of conglomerate, breccia, shale, siltstone, mudstone, sandstone, graywacke, and grit. The breccia and conglomerate occur in thick nonpersistent bands. At Colfax, the Mariposa breccia rests on the Clipper Gap formation with a depositional unconformity.

Lawson (Moody, 1917, p. 384) advocated a glacial origin of the Mariposa breccia, especially the breccia exposed at Colfax, and thought it was a tillite. Moody (1917) suggested the possibility of transport of materials by two sets of streams: turbulent streams occupying gorges cutting across the ridges and gentle subsequent streams. According to him, the first set of streams probably brought in coarse material, while the second set carried finer silts. The writer has noted that the breccia is made up of irregular masses of shale, slate, phyllite, sandstone, grit, and conglomerate. Some of these rock types are so similar to those of the Clipper Gap formation in the immediate neighborhood that they appear to have

Table 2. *Chemical analysis of volcanic rock from Logtown Ridge formation compared with other volcanic rocks.*

Constituent	1	2	3	4	5	6
SiO_2	51.26	51.22	51.35	53.30	51.01	49.24
Al_2O_3	15.10	13.66	16.34	15.16	11.89	14.79
FeO	9.06	9.20	6.19	8.71	6.08	8.00
Fe_2O_3	1.07	2.84	4.64	2.54	1.57	1.52
TiO_2	0.98	3.32	2.74	2.41	0.98	0.96
MnO	0.17	0.25	0.20	0.28	..	0.18
CaO	8.32	6.89	6.61	2.97	10.36	10.74
MgO	5.92	4.55	3.73	4.14	8.87	6.89
K_2O	0.25	0.75	1.94	0.32	0.15	0.88
Na_2O	4.64	4.93	5.01	5.55	4.17	2.76
H_2O- 105°C	0.39	1.88	--	0.18	0.24	0.02
H_2O+ 105°C	2.72	--	--	3.14	2.09	2.97
CO_2	0.28	0.94	--	nil	..	0.90
P_2O_5	0.07	0.29	1.00	0.51	0.17	0.17
Rest	--	--	--	0.40	--	--
Total	100.23	100.72	99.75	99.61	97.58	100.02

Analyst: W. H. Herdsman.
1. A flow in the Logtown Ridge formation, Ponderosa Way, NW ¼ of Sec. 1, T. 13 N., R. 9 E., Colfax quadrangle.
2. Average spilite (Sundius, N. Geol. Mag., vol. 67, p. 9, 1930).
3. Average oligoclase andesite of Hawaiian Islands (Macdonald, G. A., Div. Hydr. Terr. Hawaii Bull. 11, p. 102, 1947).
4. Albite diabase, Poorman Mine, Oregon (Gilluly, J., American Jour. Sci., vol. 29, p. 235, no. 6. 1935).
5. Diabase from Grass Valley, Nevada County, U.S. Geol. Survey, 17th An. Rep., pt. 2, p. 150, 1895-96.
6. Diabase dike, west of Jackson, Amador County. An augite porphyrite with abundant augite phenocrysts. U.S. Geol. Survey, 14th An. Rep., pt. 2, 1892-93, p. 473.

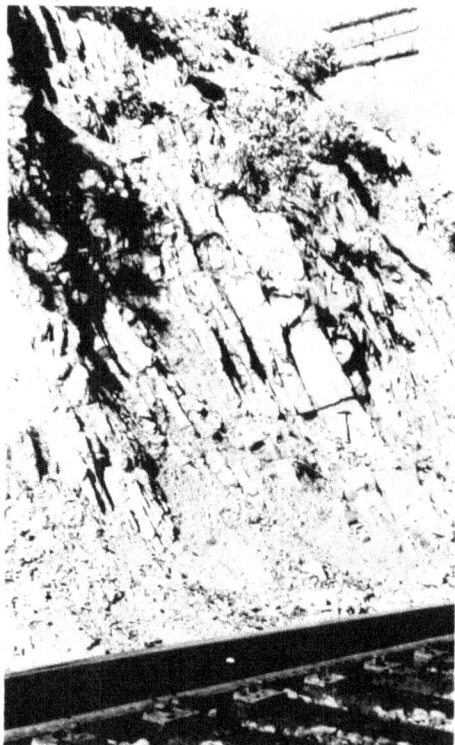

Photo 9. Thin-bedded chert and intercalated slate, Clipper Gap formation, in railroad cut at Weimar.

Photo 10. Clipper Gap formation: conglomerate containing pebbles of shale and sandstone in a shaly matrix. On Cape Horn road. The scale is 6 inches long.

been derived from the latter formation by submarine landslides at the beginning of deposition of the Mariposa formation. Crowell and Winterer (1953) have explained the formation of a similar type of breccia in the Knoxville formation, in the Chico group, and in the Pico formation. In their opinion the "beds of graded conglomerate, laid down on soft water-saturated mud, became unstable and slumped down-slope, mixing the pebbles with the mud."

Conglomerate beds in the Mariposa formation often contain lenses of grit, mudstone, and sandstone. Such alternation of fine and coarse sediments may be accounted for by assuming "pulsations in the supply" (Kuenan and Menard, 1952).

Conglomerate is composed of angular, subangular, and rounded pebbles, cobbles, and boulders of chert, quartz, quartzite, meta-andesite, slate, sandstone, and graywacke. A thin section of conglomerate contained pebbles of dolomite, chert, basalt, diabase, slate, graywacke, and fragments of quartz and feldspars, in a matrix of tuff and clay.

Graywacke is dark-colored, fine- to medium-grained, and usually shows graded bedding. It consists of angular and subangular grains of quartz, orthoclase, plagioclase, epidote, and iron ore in a clay matrix. One of the thin sections shows abundant augite and hornblende besides the usual constituents already mentioned. Evidently the rock contains some basic tuff.

Lindgren (1900) reported occurrence of fossils in the Mariposa formation at two places: at the first place one ammonite, *Perisphinctes colfaxi* Gabb and at the other place specimens of the ammonite *Olcostephanus lindgreni* Hyatt. In his opinion, "the fossils point to the uppermost Jura-Trias." Dr. Howel Williams has reported some ammonites but the details are not available. From fossil evidences elsewhere Taliaferro (1943, p. 284) concludes that the Mariposa is Oxfordian and, possibly, lower Kimmeridgian (both stages belonging to the upper Jurassic epoch).

Pre-Nevadan Intrusive Bodies

Intrusive bodies that invaded the bedrocks after or during the later part of deposition of the Mariposa formation, but before the Nevadan orogeny, are mainly diabasic and andesitic. Felspar porphyries are common and often form the marginal facies of diabase masses. There are also several spilitic bodies. These intrusions have been folded along with their host rocks during the Nevadan orogeny.

Diabase. Diabase is widely distributed in the Clipper Gap and Mariposa formations as sills, dikes, and rarely as flows. The diabase sill in the Clipper Gap formation on

Photo 11 (right). Steep-dipping Mariposa conglomerate at Ponderosa road bridge on North Fork of the American River.

Photo 12 (left). Bedded Mariposa conglomerate at same locality as photo 11.

the west side of the Bear River is at least 1500 feet thick. In some places along its eastern border it shows a slightly discordant relationship to the host rocks. There the diabase resembles a flow-breccia containing fragments of the country rock. Immediately west of Colfax, there is another diabase body in which a small gabbro body has been emplaced. The contact, wherever exposed, is well defined, and does not indicate that the diabase is the chilled phase of the gabbro, but instead is a separate intrusion. There are other minor diabase bodies intrusive into the Clipper Gap and Mariposa formations.

The diabase is normally medium-grained but occasionally fine-grained and basaltic in appearance. The texture is essentially diabasic. Where the rock contains phenocrysts of augite, it shows locally an ophitic texture. In two thin sections, plagioclase microlites in the groundmass show a trachytic texture around the phenocrysts of plagioclase and augite. Amygdules are rare, except in rocks close to the margins of intrusions.

The diabase is composed chiefly of plagioclase and pyroxene. The plagioclase, where unaltered, is labradorite (An_{61-63}) and the pyroxene is augite ($2V = 55°$; $Z \wedge c = 43°$). Plagioclase retains its euhedral outline, and is usually zoned and twinned. Augite is often pseudomorphically replaced by calcite and/or chlorite. The groundmass is composed of fine plagioclase laths, augite, iron ore, and alteration products such as epidote, chlorite, calcite, and titanite. Thin sections of flow-breccia contain angular fragments of slate, chert, and fine or glassy basalt. Amygdules, if present, are composed of calcite or chalcedony.

A thin section of diabase taken from a mass intrusive into the Mariposa formation and terminated at both ends by serpentine, contains augite ($2V = 38°$ to $41°$) somewhat more magnesian than typical diopsidic augite. Presence of abundant chlorite and some antigorite in the rock may indicate that serpentine intruded the diabase and hence is younger than the diabase.

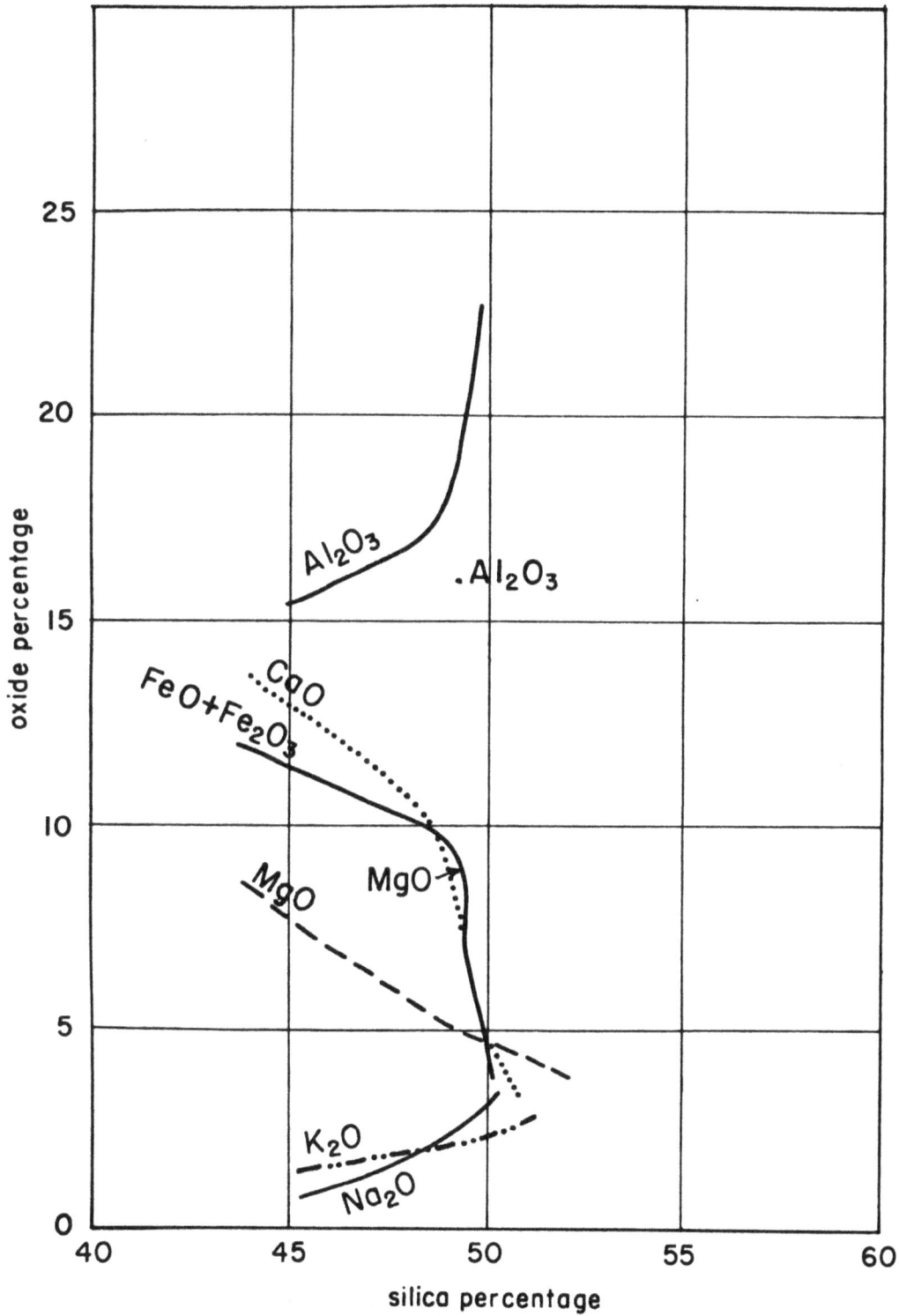

VARIATION DIAGRAM
Basic volcanic rocks of Cape Horn formation
(Analyses 1 to 5 in table 1.)

FIGURE 2.

Albite Diabase. Most of the minor intrusive diabase bodies in the Mariposa and Clipper Gap formations, especially those bodies which are near serpentine intrusions, are of albite diabase. The diabase sill intrusive into the Mariposa formation in the Codfish Creek area and the small plug-like intrusion at Coyote Hill in the Clipper Gap formation are among them. These two bodies, intrusive into conglomerate beds, contain as xenoliths numerous pebbles of slate, chert, graywacke, sandstone, and older basic rocks, evidently derived from the wall rock. Some of these pebbles have indurated borders.

The extremely fine grain of the albite diabase is similar to the diabase described earlier. The mineral constituents in the two types are almost the same, the important difference lying in the composition of plagioclase. In the albite diabase the plagioclase is invariably albite (An_{0-10}, the majority of crystals being An_5). The high 2V (72°–82°) and the positive optic signs show that the albite has formed below 700° C and so is probably a secondary, and not a primary igneous mineral (Tuttle and Bowen, 1950). A few of the rocks resemble flow breccia containing fragments of fine diabase, glassy basalt, or devitrified glass with spherulites but the identity is uncertain owing to weathering and metamorphism. One of the thin sections contains glass replaced by nontronite. Diabase adjoining serpentine bodies often contains antigorite and prehnite in the groundmass; the prehnite is also present as thin veins.

Feldspar-Porphyry. Feldspar porphyry occurs as thin sills and dikes, and sometimes forms marginal parts of larger diabase bodies in the Mariposa and Clipper Gap formations.

Feldspar-porphyry is medium- to coarse-grained and is distinctly porphyritic with large phenocrysts of plagioclase. The phenocrysts are zoned and twinned, but are invariably altered. The groundmass is composed of plagioclase (usually saussuritized), augite (replaced by chlorite), and secondary products such as actinolite, chlorite, epidote, titanite, leucoxene, and carbonate minerals. Though the rock commonly is hydrothermally altered, the texture is well preserved.

Andesite. Andesite intrusions are few in number and occur as thin sills in the Clipper Gap and Mariposa formations. Rocks from two intrusive bodies were studied under the miscroscope; one of them occurs along the Foresthill road, while the other is exposed along Highway U.S. 40 at the north margin of the map.

The andesite is massive and medium-grained. It is essentially porphyritic with phenocrysts of plagioclase (An_{36}) and augite (partially replaced by chlorite) in a groundmass composed of plagioclase microlites, augite, ilmenite, sulphide, and secondary products: chlorite, epidote, calcite, and titanite. One of the thin sections contains some quartz.

Photo 13. Trace of bedding planes on weathered joint surface, Mariposa formation, Foresthill road. The scale is 6 inches long.

Spilite. Several thin intrusive bodies of spilitic rock have been noted in the western part of the area. They are confined to a fault zone between the Clipper Gap and Mariposa formations, and are generally associated with intrusive serpentine bodies.

Spilite is a fine-grained rock. The texture is typically basaltic. It is composed of albite (An_{0-10}, usually An_5), augite (2V = 54°; $Z \wedge c$ = 43°) though rarely preserved, iron ore, and secondary products: actinolite, chlorite, epidote, calcite, titanite, and limonite. Augite is usually replaced by actinolite, chlorite, and calcite. Prehnite is present as veinlets; it also replaces plagioclase and fills up the amygdules along with calcite. In one of the thin sections pumpellyite (with a small + 2V) is abundant.

Nevadan Intrusive Bodies

Nevadan orogeny was accompanied by many intrusions —ultrabasic rocks, gabbro and granodiorite, quartz veins, lamprophyre, and keratophyre.

Ultrabasic Intrusive Bodies

Numerous bodies of ultrabasic rocks are intrusive into the Mariposa and Calaveras formations. Most of them are extensively serpentinized, the original rocks being rarely preserved. Larger bodies of serpentine contain some cores of massive rock in foliated serpentine. Study of these cores indicates that most of the intrusions were of dunite, some were of peridotite and a few of pyroxenite.

Dunite. Unaltered dunite was not noted anywhere in the mapped area. More than half of the thin sections made from massive cores in serpentine bodies are of partially altered dunite. These dunite cores are distributed in the large serpentine body occupying the northeast corner of the area. The shape of the ultrabasic intrusion is not apparent; it might be a very thick sill or a large irregular intrusive body.

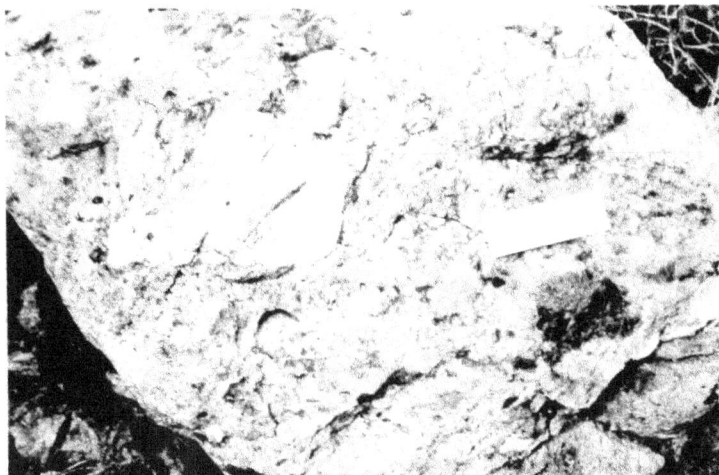

Photos 14 (above) and 15 (right). Albite diabase containing xenoliths from host rock—Mariposa conglomerate. Codfish Creek area. The scale is 6 inches long.

The massive cores are coarse grained and have a gabbroic texture. They are cut by thin serpentine veins. The rock is mainly composed of olivine and enstatite, iron ore and/or chromite. The olivine has a very large 2V and sometimes shows undulose extinction and translation lamellae due to deformation. It is partially or completely replaced by antigorite. Enstatite has been replaced by bastite which, in turn, has been replaced by antigorite. Some chlorite is usually present, surrounding the chromite or iron ore. Part of the iron ore is derived from olivine. Antigorite is occasionally replaced by talc. One of the thin sections shows complete pseudomorphic replacement of enstatite by talc.

Peridotite. Peridotite occurs as a thin sill in the Cape Horn formation southwest of Windy Point and is exposed along the North Fork of the American River. Several massive cores in serpentine are of altered peridotite.

The peridotite is a massive coarse grained rock having a gabbroic texture. It consists of olivine with deformation lamellae, augite with medium 2V and strong dispersion, bytownite, iron ore, and secondary products such as actinolite, titanite, leucoxene, chlorite, and antigorite. The mafic minerals are partially replaced by antigorite.

Pyroxenite. A thin lens of pyroxenite occurs in a serpentine body exposed about a mile north of Weimar, along the west-bound railway track. Another outcrop of partially altered pyroxenite is in the form of a sill in the Cape Horn formation along the North Fork of the American River, southwest of Windy Point. One of the massive cores in the serpentine is altered pyroxenite.

Pyroxenite is a massive coarse-grained rock with a gabbroic texture. It is cut by thin serpentine veins. The rock consists chiefly of augite ($2V = 52°$; $V \wedge c = 41°$) and enstatite. Augite is partially replaced by antigorite and talc. In one thin section, enstatite shows lamellar structure due to deformation. The enstatite is partially altered to antigorite. Prehnite is present in the form of thin veins. A thin section of massive core, mentioned before, shows an interesting replacement phenomenon: the pyroxene is replaced by bastite and the latter in its turn by carbonate minerals along cleavages.

Hornblendite. A small body of hornblendite occurs as an inclusion in serpentine south of Burnt Flat.

The rock is coarse grained and consists chiefly of hornblende which is replaced by actinolite along borders. The other minerals constituting the rock are actinolite, chlorite and talc (replacing actinolite), titanite and iron ore.

Serpentine. Serpentine is widely distributed in the area, and is usually present along fault zones. It occurs as sills and occasionally as irregular bodies showing a discordant relation to the country rock. It is often associated with masses of ultrabasic rock.

Two types of serpentine may be distinguished on the basis of megascopic appearance. They are, however, so intimately associated with one another that they cannot be mapped separately. The first type is dark green and massive and is chiefly composed of bastite or antigorite with some minor amounts of chromite and/or iron ore. The mafic minerals, olivine and pyroxene, are pseudomorphically replaced by bastite or antigorite; but occasionally the replacement is partial, leaving some cores of mother minerals. Iron ore may be primary or secondary. Serpophite is sometimes present. Bastite, in its turn, is partially or completely replaced by carbonate.

The second type of serpentine is the sheared type which is strongly foliated and is traversed by slickenside surfaces. This rock is often cut by veinlets of asbestos.

Basic and ultrabasic rocks closely associated with serpentine intrusions have been subjected to one or more of the following changes: replacement of some or all of the mafic minerals by bastite, antigorite, talc, and/or a carbonate mineral, albitisation of plagioclase (soda metasomatism), replacement of plagioclase by prehnite (lime metasomatism), formation of pumpellyite (soda metasomatism)—the last one being rather rare.

Field and petrographic evidences suggest that the serpentine is slightly younger than the gabbro.

Gabbro

A small boss of gabbro was emplaced in diabase west of Colfax. The contacts between gabbro and diabase, wherever exposed, are well-defined, indicating that they have been intruded separately. About a mile north of Applegate, a small body of gabbro is intrusive into the Clipper Gap formation. In a few other places, still smaller bodies of gabbro are exposed near serpentine intrusions.

The gabbro is coarse-grained with a typical gabbroic texture. The rock consists of diallage with uralitised borders, plagioclase (invariably saussuritized), hornblende (partially replaced by actinolite), titanite, epidote, clinozoisite, chlorite, iron ore, and sulphides. Rocks adjoining serpentine bodies contain abundant chlorite that has partially replaced mafic minerals, and prehnite developed at the expense of plagioclase; these rocks are also cut by thin veins of chlorite. A thin section of pegmatitic hornblende-gabbro shows typical poikilitic texture—hornblende crystals enclosing plagioclase laths.

Granodiorite

West of Colfax, granodiorite occurs as a small body probably intrusive into gabbro but the contact between gabbro and granodiorite is not exposed. In one place the granodiorite has a sharp contact with diabase in which the gabbro body is emplaced. The northern and western border zones of the granodiorite body are rich in mineralized quartz veins and stringers.

The granodiorite is medium- to coarse-grained. It has an essentially granitic texture, but constituent minerals show effects of brecciation. The rock is composed of quartz, orthoclase, myrmekite, oligoclase (mostly albiclase and some albite ($An_{2.5}$), hornblende invariably replaced by chlorite and iron ore), ilmenite, and alteration products such as epidote, chlorite, leucoxene, iron ore and calcite. Pyrite is finely disseminated in the rock whereas calcite is present as thin veins. Quartz shows strain-polarization; and feldspars are granulated along borders. The granodiorite along the northern and western

Photo 16 (left). Lamprophyre dike showing well defined contact with the slate of the Blue Canyon formation. Mosquito Ridge road.

Photo 17 (below). Lamprophyre dikes in the Blue Canyon formation. Mosquito Ridge road.

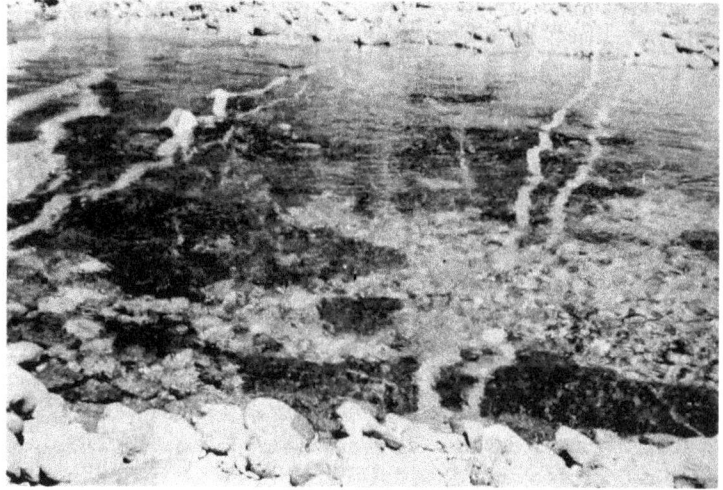

Photo 18 (left). Lamprophyre dike in metabasalt, Cape Horn formation, on North Fork of the American River. The scale is 6 inches long.

Photo 19 (above). Keratophyre dikes in Cape Horn slate, across bed of the North Fork, American River.

border shows greater microbrecciation than in the central part.

Quartz Veins

Quartz veins are usually thin, lenticular and often steep dipping. They are more abundant in the Mariposa formation than in any other. Many of these veins are auriferous and have been profitably mined in the past.

Lamprophyres

Lamprophyres occur as narrow dikes in the Blue Canyon and Cape Horn formations. They are abundant in the southeast corner of the area. Sharp, chilled contacts with the country rocks are characteristic. Almost all the lamprophyres are kersantite.

The lamprophyre is a dark-colored, porphyritic dike rock rich in mafic minerals. It is composed of abundant biotite, amphibole (usually barkevikite), andesine (An_{42-48}), titanite, and iron ore. Actinolite, chlorite, epidote, clinozoisite, calcite, and rutile are the secondary minerals in the rock. Biotite is often bleached or replaced by chlorite; barkevikite or hornblende is altered to actinolite; plagioclase is sometimes saussuritized. Presence of quartz in some thin sections may be due to contamination; a part of the quartz is probably secondary. Rutile is formed at the expense of biotite. The lamprophyres are often carbonatised and their porphyritic texture is often replaced by a sugary texture.

Keratophyre

Keratophyre occurs as thin sills and dikes. Some of them are intrusive into the Cape Horn formation and are exposed along the North Fork of the American River. Two other bodies are emplaced in serpentine.

The keratophyre is a massive, light-colored rock. It is fine-grained in the narrow intrusions, but medium- to coarse-grained in the thicker bodies. Texture is porphyritic, but often aplitic. The fine-grained rock consists of albite (An_{0-5}), actinolite, chlorite, epidote, titanite, quartz, iron ore, and sulphides—pyrite and chalcopyrite. The porphyritic keratophyre contains phenocrysts of albite and altered mafic minerals in a groundmass of albite, chlorite, epidote, titanite, and iron ore. The phenocrysts are often granulated along edges. In coarse-grained keratophyre, the albite is often altered to clay mineral. Carbonate minerals are present in the form of thin veins.

Felsite

Felsite occurs as a dike in the Clipper Gap formation and is exposed across the Bear River. It is massive and well jointed.

Felsite is pinkish red, rather fine grained, and sugary textured. It consists chiefly of albite (An_5) and limonite. Albite is the predominant mineral. Limonite is widely distributed and might be secondary after mafic minerals, sulphides, or even iron-oxide. The minor constituent minerals in the rock are chlorite, sericite, and quartz.

Other Rock Types

The following metamorphic rocks, though sparsely distributed, have been found as inclusions in or in close association with serpentine bodies.

Tremolite-schist is composed chiefly of tremolite partially replaced by antigorite and chlorite or by talc. Iron ore, titanite, and epidote are the other constituent minerals.

Chlorite-rutile rock consists largely of chlorite with some rutile.

Talc rock is distributed in the marginal parts or shear zones of serpentine bodies. A thin section of talc rock shows talc, chlorite, and titanite.

Albite-schist occurs as a massive inclusion in the serpentine body east of Foresthill divide. The rock is fine to medium grained and has a sugary texture. It is composed of albite (An_{0-4}), actinolite, chlorite, epidote, clinozoisite, quartz, biotite (bleached), and titanite.

Metasomatised volcanic breccia is exposed at the Old Vore Mine in the Codfish Creek area. It is associated with a thin serpentine intrusion. The rock is fine grained and has a foliated appearance. It consists of albite, chlorite, quartz, titanite, iron ore, calcite, and stilpnomelane. It also contains fragments of igneous rock and shale. The thin section shows some shear lines and thin calcite-chlorite veins. Stilpnomelane is extensively developed, indicating a low grade metamorphism.

Carbonate rock near Mosquito Ridge road is present as a thin ankerite vein trending almost north-south in a serpentine body. This vein is associated with stringers of quartz that often contain gold. Formation of these quartz stringers may be owing to release and concentration of silica from immediately adjacent country rock which is converted into ankerite by carbon dioxide (Knopf, 1929).

In the Codfish Creek area, serpentine is locally replaced by carbonate minerals. Replacement of antigorite by carbonate minerals has been noted in several thin sections of serpentine.

Auriferous Gravels

Auriferous gravels which do not contain fragments of rhyolite belonging to the Valley Springs formation or to any earlier Tertiary formation will be discussed under this heading. Such auriferous gravels which are of the same age as the Ione formation are found in almost all placer deposits within the area and will be discussed later. The three units recognized, in order of decreasing age are: carbonaceous shale, bench gravels, deep gravels.

Deep gravels filled the channel bottoms. They contain pebbles, cobbles, and boulders of older rocks, all fairly rounded or subrounded, indicating long transportation from the source area. They are often cemented and generally carry more gold than other gravels.

Bench gravels which constitute an important member of the Ione formation elsewhere, are exposed in a few placer mines within the present area. They contain rounded or subrounded pebbles, cobbles, and boulders, often loosely cemented, and are intercalated with non-

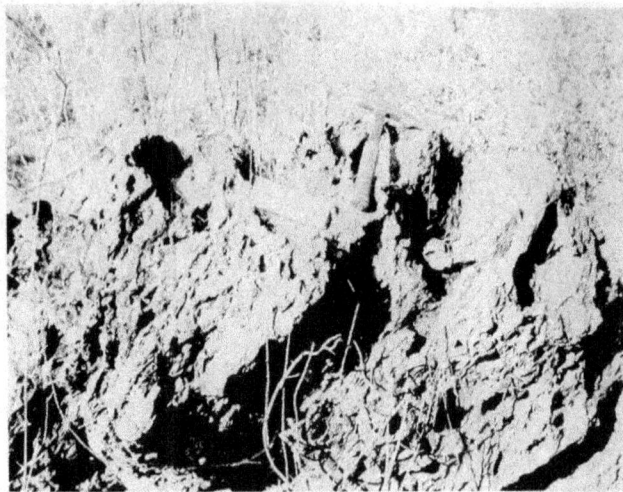

Photo 20. Carbonate rock along the marginal part of serpentine body in the Codfish Creek area. The pick and 6-inch ruler are for scale.

persistent sandy shale and sands which show false bedding. The bench gravels have been worked for gold, but they are poorer in gold content than the deep gravels.

Carbonaceous shale, intercalated in the bench gravels, has been found in one place—Elizabeth Hill Mine. The shale, though less than a foot in thickness, resembles similar Ione members exposed near the town of Ione (the type locality).

Beds resembling those of the Ione formation have been noted in two mines, the Buckeye (in sec. 14, T. 14 N., R. 10 E.) and Elizabeth Hill (in Sec. 5, T. 14 N., R. 10 E.). In other mines the bench gravels are either thin or missing and shale has not been detected. There are several places where a thin gravel bed presumably belonging to the "auriferous gravels" grades upward into gravels of the Valley Springs formation. The distinguishing feature of the latter formation is the presence in the gravels of fragments of rhyolite. The pebbles (other than those of rhyolite) forming the gravel beds of the Valley Springs formation are exactly similar to those of the auriferous bench gravels and presumably were derived from the latter as a result of reworking by intervolcanic streams. Stratigraphically, such reworked gravels belong to the Valley Springs formation, since they were laid down during the intervolcanic period. The hypothesis that the Ione deposits were extensively eroded by intervolcanic streams may explain the absence of second and third members of the "auriferous gravels" in most of the placer deposits in the area.

Fossils collected in the past from the auriferous gravels and the overlying beds (the Valley Springs formation) in Iowa Hill locality (Lindgren, 1911, pp. 54-64) range in age from lower Eocene to upper Miocene (Chaney, 1932). MacGinitie (1941) reported fossils from the flood-plain deposits interbedded in the auriferous gravels in

Photo 21. Bench gravels resting on pre-Ione surface at Gleeson Diggings.

Photo 22. Rhyolite gravel bed at Gleeson Diggings.

Chalk Bluffs locality 8 miles north of the present area, and concluded that the age of the gravels was lower middle Eocene and that the gravels and fossil-bearing sediments were equivalent to the beds of the Ione formation. Potbury (1937) found fossils near La Porte (about 30 miles northwest of Chalk Bluffs) in hypersthene dacite vitric tuff unconformably overlying carbonaceous shales and auriferous gravels, and concluded that the leaf-bearing tuff was not older than late Eocene in age. From the lithology and stratigraphic position of the carbonaceous shale at the Elizabeth Hill Mine, the writer considers that the "auriferous gravels" of the present area are of the same age as the auriferous gravels of Chalk Bluffs and La Porte areas and that these gravels are stratigraphically equivalent to the Ione formation of the type locality.

Valley Springs Formation *

Sands and gravels overlying the "auriferous gravels" and underlying the Mehrten formation, and containing pebbles or fragments of rhyolite of Tertiary age, have been grouped under the Valley Springs formation. The upper part of this formation consists of rhyolite tuff.

The Valley Springs formation is well distributed in the area, and varies from a few feet to 400 feet in thickness. The formation is nearly flat, dipping west or southwest at 0° to 5°. It consists of interbedded gravels, sands, and tuffs. The gravel beds contain abundant pebbles of rhyolite and are intercalated with lenses of sands containing abundant biotite. The sands are predominantly rhyolite ash mixed with other sediments. There are also boulder beds containing abundant rhyolite boulders along with boulders of other rock types. The rhyolite sands and

---

* The author, in his Ph.D. dissertation, introduced the new name "Rhyolite formation" for this unit. As the name is not acceptable, for several reasons, the editors have substituted the approximately equivalent name Valley Springs, already established in the literature on the Sierra Nevada.

tuffs are usually current bedded and appear to be fluviatile or shallow-water deposits. The rhyolite tuffs forming the upper part of the formation are compact.

The pebbles, cobbles, and boulders constituting the gravel beds of the Valley Springs formation resemble those of the "auriferous gravels", the main difference being the presence of rhyolite gravels in the Valley Springs formation. This may indicate that the "auriferous gravels" were reworked by intervolcanic streams carrying rhyolite detritus. Like the Eocene streams, the intervolcanic streams were sluggish and overloaded with sediment; this is evident from flood-plain deposits often half a mile wide across the channel.

Stratigraphic Sections at Some Placer Mines

In some placer mines, good stratigraphic sections are exposed in the scarp.

Adam's Pit: In N½ sec. 27, T. 14 N., R. 10 E.

Mehrten formation	{	mudflow
		fine andesite ash
		gravel bed
		andesite boulders

slight disconformity

Valley Springs formation	{	rhyolite gravels
		rhyolite sands and tuffs
		rhyolite gravels

unconformity

bedrock

The Mehrten formation overlies the Valley Springs formation with a slight disconformity indicating a time interval between the two formations. In other places the two formations are conformable.

Big Dipper Mine.—Sec. 3 and 4, T. 14 N., R. 10 E. A thick sequence of gravels resembling those of the "auriferous gravels", but containing a few rhyolite pebbles, is underlain by waterlaid rhyolite tuff.

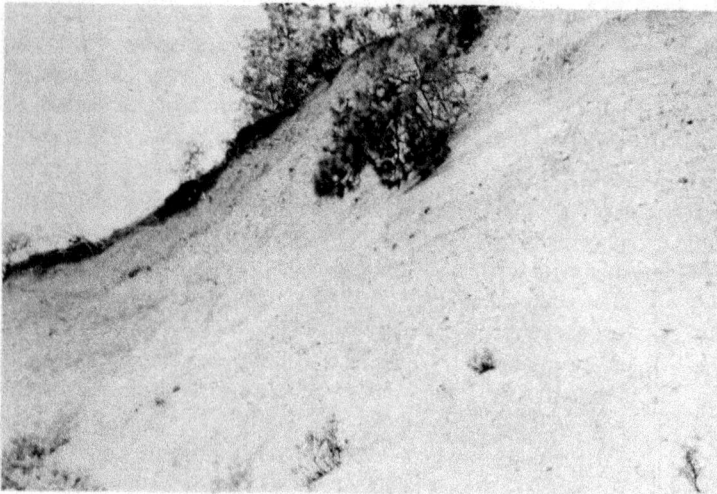

Photo 23.   Rhyolite sand and gravel underlying a mudflow (seen at the skyline). First Sugar Loaf.

Photo 24.   Rhyolite gravel intercalated with rhyolite sand, at San Francisco mine.

Mayflower Mine. Sec. 23, T. 14 E., R. 10 E. In the hydraulic pit, about 100 feet of the Valley Springs formation is exposed, underlying the Mehrten formation. The Valley Springs formation consists of rhyolite boulder beds, rhyolite sands (current bedded), rhyolite tuffs (in places, compact but not welded), and gravel beds with abundant rhyolite gravels.

Roach Hill. Sec. 27 and 28, T. 15 N., R. 10 E. At the portals of several adits, gravel beds containing about 10 percent of rhyolite lie directly on the bedrock. Gravel beds and rhyolite sands predominate at the base of the Valley Springs formation, while thick and massive rhyolite tuffs form the upper part. A red-colored mudflow resting on cream-colored rhyolite tuff is conspicuous in the scarp from a distance.

Placer Diggings, Southwest of Foresthill. Sec. 34 and 35, T. 14 N., R. 10 E., Sec. 3, T. 13 N., R. 10 E. A generalized sequence of the Valley Springs formation may be stated as follows:

Mehrten formation

Valley Springs formation
{
Rhyolite tuff
Rhyolite tuff (compact)
Rhyolite tuff
Rhyolite sands and gravels
Rhyolite gravels and boulders
}

Unconformity

Auriferous gravels or bedrock

The gravels of the Valley Springs formation contain up to 20 percent of rhyolite. In the scarp of the placer dig-

gings, no disconformity was noted between the Valley Springs and Mehrten formations.

Rhyolite Tuffs

Rhyolite tuffs constitute the upper part of the Valley Springs formation, and are well exposed in the scarp of most of the placer mines, especially in Roach Hill and southwest of Foresthill. The tuffs are often intercalated with cross-bedded rhyolite sands, indicating that they are of fluviatile or shallow-water deposition.

The rhyolite tuff at the Mayflower Mine is fine grained and compact, but microscopic examination indicates that it is not welded. The rock consists of phenocrysts of sanidine, oligoclase, quartz, and biotite in a matrix of abundant glass shards. Some of the shards are colorless, whereas others are pale greenish owing to slight marginal alteration to beidellite. A thin section of similar rock from almost the same horizon shows hydrothermal alteration. The green biotite is altered to chlorite along its edges, some feldspars are turbid due to kaolinization, and the colorless or pale greenish glass is more extensively altered to beidellite. The rhyolite tuff from the western scarp of Roach Hill resembles those just described.

Age of the Valley Springs Formation

Fossils collected from the "auriferous gravels" and the overlying beds of the Valley Springs formation in the Iowa Hill locality (Lindgren, 1911, pp. 54-64), are reported to range from early Eocene to late Miocene in age (Chaney, 1932). The age of the "auriferous gravels" is lower middle Eocene as determined in the Chalk Bluffs area (MacGinitie, 1941). Potbury (1937) reported fossils near La Porte (about 30 miles northwest of the Chalk Bluffs) in hypersthene dacite vitric tuff unconformably overlying carbonaceous shales and auriferous gravels; in her opinion the fossils indicate that the tuff is not older than late Eocene. The dacite tuff probably represents the beginning of volcanic activity which later culminated in the deposition of rhyolite tuffs. On this supposition, the Valley Springs formation probably is not older than late Eocene. As regards its upper age limit, it is not younger than late Miocene (Chaney, 1932).

Mehrten Formation

The Mehrten formation covers a large area and forms flat-topped ridges. The presence of scattered outcrops in the western half of the area probably indicates that the formation once covered practically the whole area but was later removed in many places by erosion. The

Photo 25. Mudflow (at the treeline) resting on rhyolite tuff and sand, at Roach Hill.

Mehrten formation conformably overlies the Valley Springs formation, but at the Adam's Pit, a slight disconformity was noted. The thickness of the Mehrten formation ranges from a few feet to about 700 feet.

The formation is composed of andesite tuffs, breccias, sands, gravels, and boulder beds. Andesite is by far the most predominant rock type in the beds of the formation, but occasionally pebbles and boulders of meta-andesite, serpentine, granodiorite, quartz, and rhyolite are present with andesite. No andesite flow belonging to this period (Mio-Pliocene to Middle Pliocene) has been noted in the area. Andesite breccia is the most abundant part of the formation and will be described later. Andesite sand and gravel beds represent intervolcanic fluviatile or floodplain deposits. These gravel beds are locally auriferous, the placer gold being probably derived from the "auriferous gravels."

Andesite breccia and conglomerate predominate in the Mehrten formation. Similar breccia and conglomerate in other parts of the Sierra Nevada have been identified as andesite mudflow or lahar. Its occurrence and mode of origin have been studied in detail by Curtis (1951) in the Topaz Lake quadrangle.

Andesite is the chief, if not the only, constituent forming the mudflow which consists of fragments, pebbles and boulders of andesite in a matrix of fine andesite tuff. The andesite boulders show some variation in physical appearance such as granularity, color, and vesicularity and also size and shape. The andesite breccia at the Pond's Pit is chiefly composed of scoria and is lighter in color and weight.

Several thin sections of andesite boulders constituting the mudflow were studied. The rock is medium-grained and has a porphyritic texture. It consists of phenocrysts of augite ($TV = 52°$; $Z \wedge c = 42°$) and plagioclase

Photo 26. Dark-colored mudflow resting on cream-colored rhyolite tuff and sand, at Roach Hill.

(An_{55-24}) in a groundmass of augite, plagioclase microlites, iron ore, yellowish brown glass (often devitrified), and limonite. Augite is twinned and often zoned showing hourglass structure. Phenocrysts of plagioclase are zoned and twinned. Dusty iron ore is abundantly distributed. In one of the thin sections, hornblende shows a poikilitic structure with laths of plagioclase embedded in it.

Condit (1944) reported plant fossils from "andesitic sediments equivalent to the lower Mehrten formation" in the Remington Hill on the southeast side of the Chalk Bluff Ridge, about eight miles north of the Colfax-Foresthill area. In his opinion, "a transitional Mio-Pliocene age" is indicated by the paleobotanical evidence.

Terrace Gravels

Terrace gravels are present in many places along the banks of the American River, and less commonly along

Shirttail and Bunch Canyons. They are 50 to 100 feet above the water level and have been worked for gold on a small scale locally.

Tailings

Tailings are widespread in the area. Some of the streams adjoining placer deposits have their banks made up of tailings. When reworked by streams, these tailings form rich pockets of placer gold in odd spots. In fact most of the placer gold recovered recently is from the old tailings.

Photo 27. Mudflow containing many large rhyolite boulders. Placer diggings southwest of Foresthill.

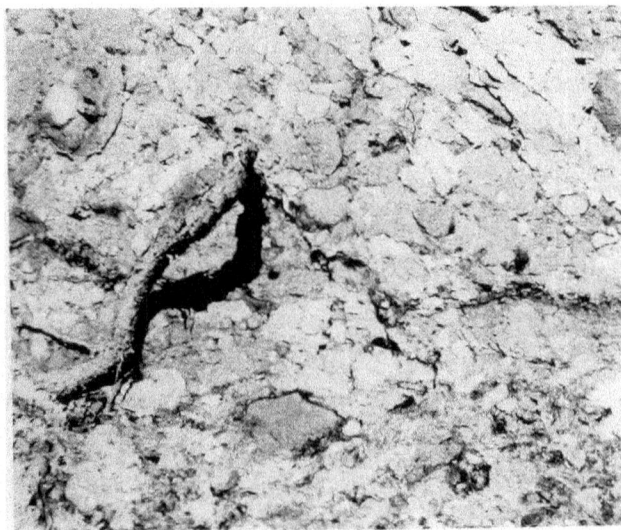

Photo 28. Close-up of a mudflow at Monona Flat.

STRUCTURE

General Large-Scale Structure. The Jurassic and pre-Jurassic rocks of the area were severely folded and faulted by the compressive forces of the Nevadan orogeny which according to current general opinion began in the late Jurassic. The writer believes that the Calaveras formations may also have been subjected to an earlier orogeny which might have taken place sometime between the Carboniferous and Jurassic periods (Cloos, 1935, and Taliaferro, 1951).* Following this hypothesis the structure resulting from the earlier orogeny would have controlled the later structure developed during the Nevadan orogeny. The Logtown Ridge and Mariposa formations are strongly folded, the folds trending generally north-northwest. The folds are rarely overturned except near the southern margin of the map. The Calaveras formations, on the other hand, being affected by two deformations, are invariably overturned into isoclinal folds with their axial planes dipping east-northeast at 60° to 90°. From this it is inferred that the force responsible for this folding (which is essentially flexure folding), was regional compression acting from the east-northeast direction.

The folds of the Calaveras formations generally have horizontal axes. Some, however, plunge either north or south, while some plunge both directions from the apex. The Mariposa formation has a regional structure parallel to that of the Calaveras formations; but where in contact with or close to the intrusive bodies that are more competent, the Mariposa beds are buckled and have folds plunging at angles greater than that of the Calaveras formations. The angle of plunge then may be as much as 40°, and in a few folds, is even steeper. Such is the case in the south where the Mariposa formation shows a change in strike from north-northwest to east-northeast. Later in this chapter the broad regional flexure-folding of the Mariposa formation is attributed to the earlier phase of the Nevadan orogeny, and that the folds plunging at moderate to high angles perhaps indicate a horizontal or diagonal movement which seems to be the later phase of the Nevadan orogeny.

There are several faults—Weimar fault, Bear River fault, and Milk Ranch thrust—between the Clipper Gap and Mariposa formations which are separated by a prominent thrust, Gillis Hill thrust, which presumably corresponds to the Mother Lode thrust farther south. The Logtown Ridge formation is bounded by the Big Bend thrust on the west and the Gillis Hill thrust on the east. The Blue Canyon formation is thrust against serpentine, the thrust being named after the Volcano Canyon. Most of these faults are parallel and have a general north-northwest trend. Some of them are closely branching.

---

* At Colfax, the Mariposa formation rests on the Clipper Gap formation with an angular unconformity.

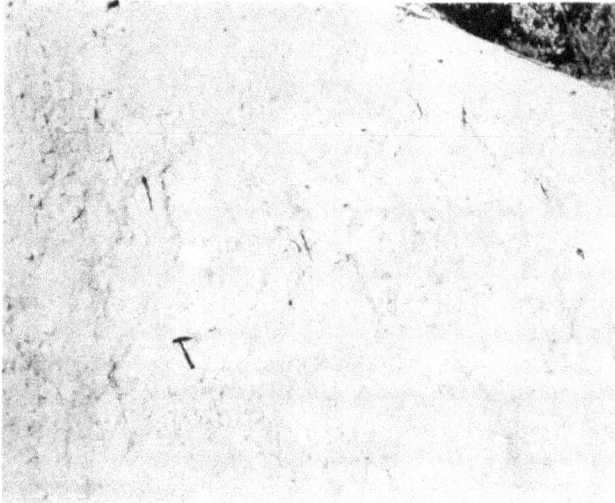

Photo 29. Mudflow showing faint bedding, Mosquito Ridge road.

They are rather steep, dipping east-northeast at 60° to 80°. Minor faults noted in the different formations, also follow the north-northwest trend shown by the major faults. The traces of some of the major fault planes show some curvature and locally form arcs. This may be partly due to the change in inclination of the fault planes, especially when in contact with more competent rocks.

In discussing the thrust-faults in the Allegheny district of Sierra County, Ferguson and Gannett (1929) suggested that "the thrust faults are the result of a compressive force acting from the east to the west, possibly due to adjustments following the intrusion of the great Sierra batholith", and that "the localization of these small thrust faults in the Allegheny district may be due to the protuberence of granite." A similar view has been expressed by Cloos (1935); in his opinion, "the direction of movement is toward the west, owing to the lateral pressure exerted by the batholith to the east." The Allegheny district is about 20 miles north of the Colfax-Foresthill area. What has been said about the former perhaps holds good for the latter.

Small-Scale Structures. The Mariposa formation often shows graded bedding, and determination of the tops and bottoms of the beds is generally rather easy. Nevertheless there are ambiguous cases where grading is reversed in adjoining laminae. Such cases are local and may be accounted for by assuming "pulsations in the supply" (Kuenen and Menard, 1952). The Clipper Gap formation presents a difficult problem since graded bedding varies sporadically. In the other three formations of the Calaveras group, intercalated volcanic ashes when not too severely metamorphosed may sometimes show faint graded bedding. Polished natural exposures in the canyon beds are helpful for the study of graded bedding in the

thin ash laminae. Although most of the canyons are rugged and have many rapids, the writer was able to revisit some of them and completed a traverse across the area mapped. Much of the detail shown in the cross sections is drawn from these observations.

Small-Scale Structure in the Cape Horn Formation. In order to elaborate the general structural picture afforded by large-scale folding, structural features (S-surfaces and lineations) developed in the rocks on a smaller scale were studied in two areas in the Cape Horn formation, which in the writer's view has been affected by the two deformations mentioned earlier.

Small-scale deformational structures observed in the Cape Horn formation belong to two categories: (1) drag folds—folds of average amplitude of 4 mm. visible in hand specimens. (The terms "drag folds" and "microfolds," have been interchangeably used.) (2) lineation defined by fine corrugations or crumpling visible under the microscope.

Oriented specimens were collected along the Shirttail Canyon road, in two areas (A and B) indicated on the index map, the first area in slate, the second in metavolcanic rock. In each area ten oriented specimens were collected at distances varying from a few feet to several hundred feet wherever there was any change in strike or dip of the foliation, which here invariably is a "bedding foliation" (Fairbairn, 1935). Laboratory examination showed that most of the specimens contain more than one lineation, and some more than one set of S-surfaces. Both lineations and S-surfaces were measured and plotted separately on equal area projections. The lineations are concentrated in a cone within 35° of the vertical, and show a slightly eccentric center of gravity (maximum in

Photo 30. Tailings stacked along banks of Indian Creek south of Monona Flat.

Diag. C, Fig. 3) about 5° north-northeast from the center of projection. Considering the small number of measurements, the two diagrams A and B (Fig. 3) are closely similar.

For microscopic examination of structure, thin sections were made from ten oriented specimens—seven of slate and three of metavolcanic rock. Two thin sections normal to the foliation were cut from each specimen—one vertical section (2-section or 4-section) and the other top or bottom section (T-section or B-section).

Drag folds are more conspicuous in the thin sections of slates than in sections of metavolcanic rocks. This reflects differences in the competence of the two rocks, the slates being less competent would allow drag folds to form more easily.

The specimens collected may be divided into two categories: those in which drag folds are consistently more obvious in 2- or 4-sections than in T- or B-sections and those in which drag folds are more conspicuous in T- or B-sections than in 2- or 4-sections. Most of the slate specimens belong to the first category. Such drag folds appear to have horizontal or gently plunging axes. Thus the attitude and style of drag folding agree with those of the major fold structures of the Calaveras formations. Several specimens of slate and most of the speciments of metavolcanic rock belong to the second category. The first category of slate specimens shows gentle crumpling of intercalated sand lenses or crumpling of shear-surfaces in their T-sections. Thus they appear to have vertical or steeply plunging folds not so conspicuous as the horizontal or gently plunging folds mentioned before. The presence of steeply plunging folds may indicate another deformation, presumably a later one. If we are correct in assuming that the Calaveras formations were already

Photo 31. Dredge tailings, Horseshoe Bar, Middle Fork, American River.

Photo 32. Gleeson Diggings, extensively mined by hydraulicking.

strongly folded during the earlier orogeny, they probably were not deformed to the same extent as the Mariposa formation was during the post-Mariposa folding. This would explain the absence of conspicuous large-scale structural features reflecting the later deformation in the Calaveras formations. On the other hand, a post-Mariposa small-scale structure such as steeply plunging lineation could develop more readily in the Calaveras rocks.

All the slates and most of the metavolcanic rocks of the Cape Horn formation show a prominent steep lineation. Tectonic interpretation of this steep lineation hinges on their alternative identity as a-lineation or b-lineation.

In general the lineation of deformed rocks falls into one of the following two categories: a-lineation is parallel to the direction of differential movement in the deformed material. It lies in the foliation (schistosity due to shear) and parallel to the symmetry plane of the fabric; b-lineation is transverse to directions of movement. It is normal to the symmetry plane of a markedly monoclinic fabric and is parallel to the axis of folding or rotation of the accompanying deformations.

The steep lineation in the slates marks the axis of microfolds which are evident in the T-sections of specimens of the second category. These microfolds agree with some large-scale deformation folds of the Mariposa sediments in the south where the folds plunge at moderate to high angles. The character of these folds gives the rock structure a monoclinic symmetry with one plane of symmetry normal to the fold axis. The lineation is thus parallel to the b-fabric axis of the steep folding; if it originated at the same time as this folding it is by definition a b-lineation (Sander, 1950). On the other hand, if the steep lineation is of the same age as the horizontal or gently plunging fold axis, it would be an

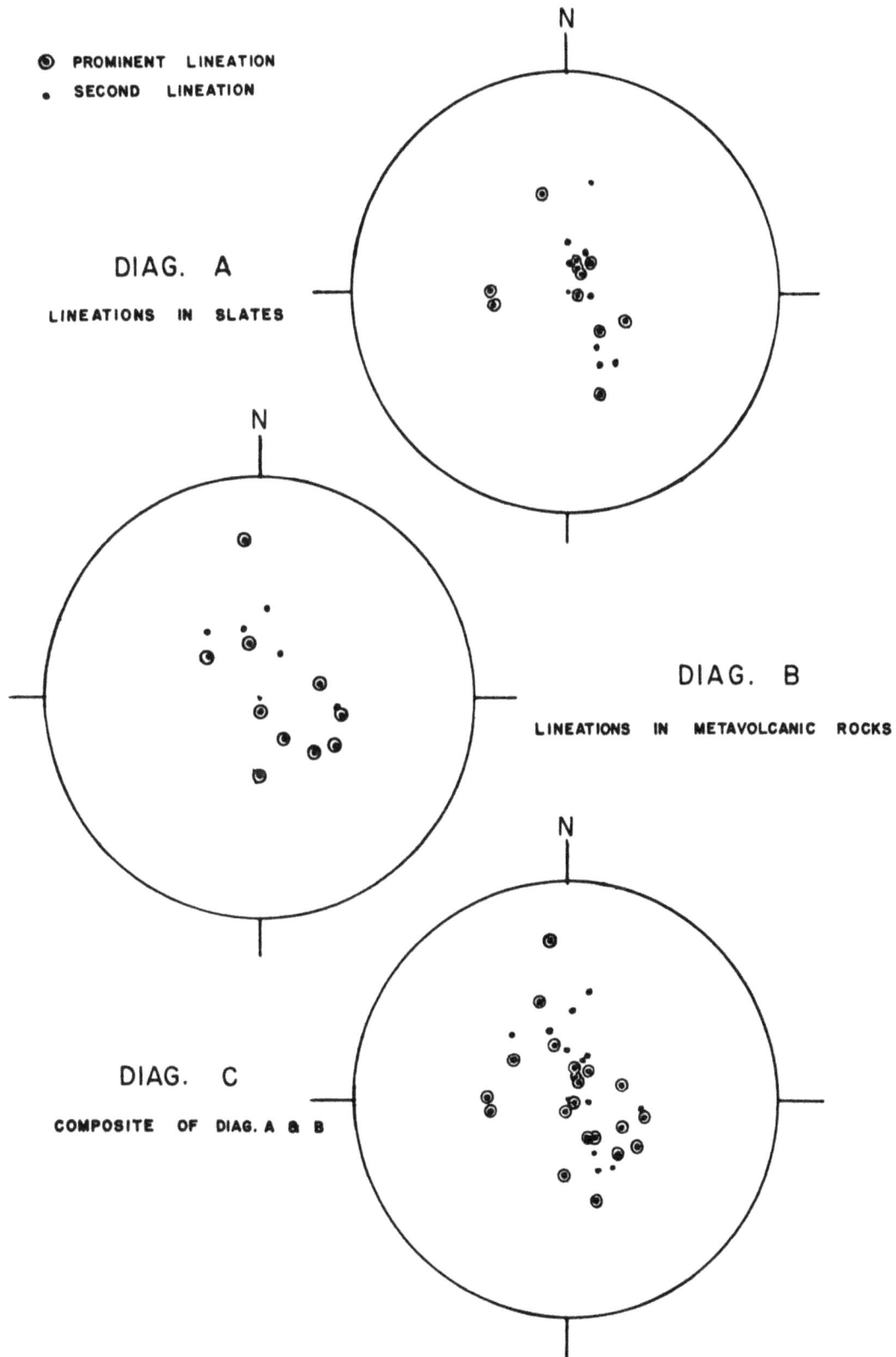

FIGURE 3.   Projections of lineations of slates and metavolcanic rocks on equal area nets.

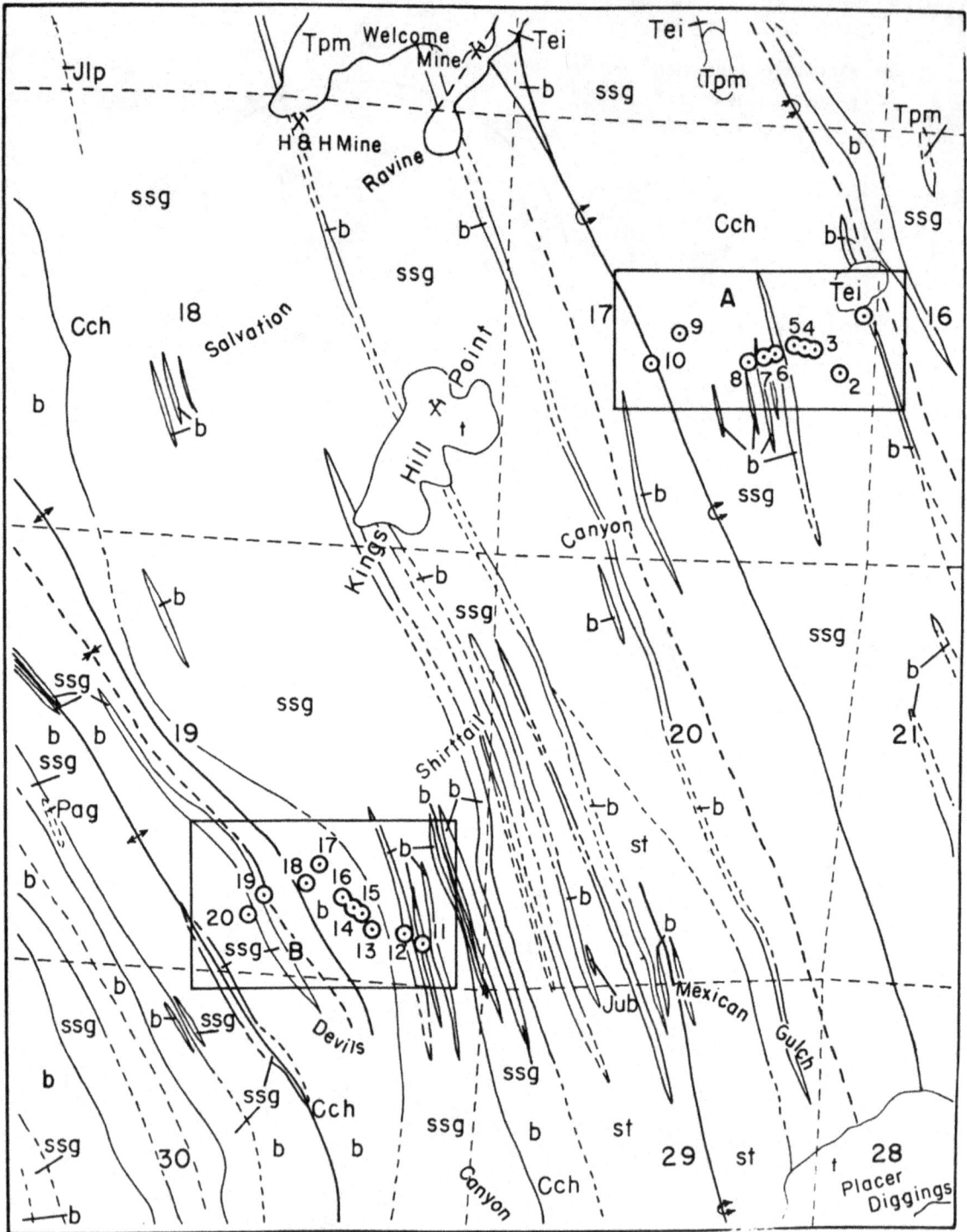

INDEX MAP SHOWING SOURCE LOCATION OF ORIENTED SPECIMENS

⊙ Location

Group A — Specimens 1 to 10 — Slates

Group B — Specimens 11 to 20 — Metavolcanic Rocks

SCALE IN FEET

FIGURE 4.

a-lineation, being then normal to the axis of contemporary folding. According to Sander and others, however, *a*-lineation is usually developed in zones of intense concentrated movement; examples occur in mylonites and slickensides. According to Cloos (1947), cleavage planes in slate might develop from shear folding with a possible development of *a*-lineation along the cleavage plane in the direction of movement, normal to the fold axis. In 2- or 4-sections none of the oriented specimens show any shear-slip of the drag-folded lenticles or laminae such as would be expected if lineation were parallel to the slip direction *a*. The drag folds show some local thickening along crests and thinning along the limbs, but no appreciable slip even on a microscopic scale. Thus the steep lineation which is the most conspicuous small-scale structural feature and is parallel to the axis of the microfolds seen in T- or B-sections, is a *b*-lineation.

No lineation corresponding to the microfolds seen in the 2- or 4-sections has been noted, though these microfolds are accompanied by large-scale deformation folds which have a horizontal or gently plunging axis. Only one specimen of metavolcanic rock shows a prominent lineation plunging almost north at 30°. The usual absence of lineation corresponding to the horizontal folds may be explained in this way: the earlier deformation folded the Calaveras rocks without producing metamorphism and lineation. *A*-lineation and schistosity only form where deformation affects all the grains, that is, where there is a strong penetrative movement (*Durchbewegung*) affecting the whole rock.

As mentioned previously, the lineations when plotted show that they are distributed in a cone of radius 30-35° from the center of projection. Hence, it may be assumed that the movement was primarily horizontal, but at times or locally was diagonal in the vertical plane, that is, the movement had a vertical component which made the folds plunge northwards and southwards. The angles at which these folds plunged depended on the competency of the folded and adjacent rocks.

Suggestion of such a late deformation due to a horizontal movement has been found in the writings of Mayo (1937) and (1941), and Locke, Billingsley, and Mayo (1940). According to Eardley (1951), "horizontal motion seems to explain many of the structural features of the regional pattern", and "the space for the intrusions" of granite was "provided mostly by buckling of isoclinally folded wall rocks as a result of north-south compression."

Structure of Calaveras and Mariposa Rocks. The Calaveras formations were subjected to two major deformations: the first after deposition of the Calaveras formations but before deposition of Jurassic formations; the second during the Nevadan orogeny. The second defor-

mation has had two phases: during the earlier phase beds underwent broad regional flexure folding under compression acting from east-northeast (the same type of compression as in the first deformation); while during the later phase, the deformed beds were subjected to a horizontal or slightly diagonal movement in an almost north-south direction. The movements in the first deformation and in the earlier phase of the second deformation might have been intense but were not strong enough to produce a lineation in the rocks. The movement in the later phase of the second deformation was penetrative and produced a strong lineation in the Calaveras rocks.

The cause of the horizontal movement in the later phase of the second deformation (Nevadan orogeny) has not been explained satisfactorily. In the opinion of some of the previous workers mentioned before, movement might be due partly to the force exerted by the emplacement of the innumerable bosses and stocks of gabbro and granodiorite and the huge batholith of granodiorite.

Considering the paucity of observations of structural features in the field, the writer cannot say more in support of the hypothesis. For a comprehensive picture, more sample areas in the Calaveras group of formations and several sample areas in the Mariposa formation should be studied. A summary of structural data is given in Table 3 (see p. 42).

Tertiary Formations. Tertiary formations overlying the bedrocks dip 0° to 5° west. A part of this dip is presumed to be due to the initial slope of the Eocene surface. The area being close to the Sierran foothills, only a minor part of the tilt, 0.5° or 1°, may be owing to the tilting of the Sierran block by the block faulting in the later part of Tertiary or early Quaternary time.

GEOMORPHOLOGY

There are three main geomorphic features in the area: the V-shaped canyons, the present flat-topped ridges capped by mudflows of the Mehrten formation, and an old Eocene surface on which the Tertiary deposits rest.

The American River, Bunch Canyon, Indian Canyon, and Shirttail Canyon cut for miles across the structural trends of the bedrocks, while the smaller streams run parallel to the trend. All the streams are in their youthful stages and have essentially V-shaped canyons. The North Fork of the American River flows for about 15 miles across the area; the features exhibited by this canyon may be considered as representative of all the canyons in the area. A panoramic view of the river shows at least two stages in canyon cutting. The first stage is shown by the upper part of the profile—a steep-sided valley with a rather flat floor—indicative of the mountain-valley stage of Matthes (1930). The second is the canyon stage indicated by the present V-shaped gorge incised in the floor of the mountain-valley stage: the walls are steep and the depth may extend to 1000 feet. The second stage is evidence of rejuvenation of the stream which started cutting into the valley floor. Downcutting is still going on as evidenced by the fact that the stream is flowing in a rock channel and has numerous rapids along its course.

The Mehrten formation is almost flat, having a very gentle slope towards the west-southwest. Near the western margin, the elevation is about 2300 feet and near the eastern end about 4000 feet. The profile of the ridges is a gently rolling surface, reduced in many places to an even tableland.

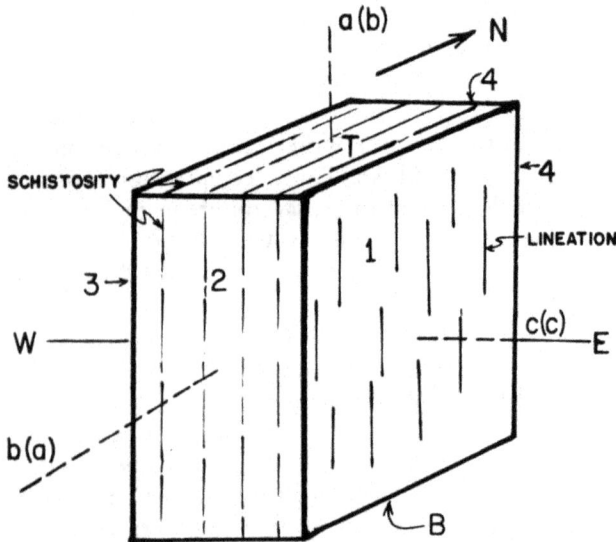

FIGURE 5. Section block of oriented specimen showing orientation of thin sections 2 or 4 and T or B with respect to geographic coordinates and fabric axes a, b, c. The fabric axes (a), (b), (c) denote the fabric axes in the last deformation.

The Tertiary formations have a gentle westerly dip. A part of the slope may be due to the tilting of the Sierran block at the time of block movement during the Pliocene and Pleistocene epochs. The extensive flood plain deposits of the Eocene and intervolcanic channels indicate a flat country in which the streams flowed sluggishly and formed meanders. Locally these streams had little gradient.

GEOLOGIC HISTORY

The oldest rocks in the area belong to the Calaveras group which consists of four formations. The lower age limit of the oldest rocks is not known. Fossils found in the uppermost member of the Calaveras group indicate a Carboniferous age.

During the Paleozoic era, at least during its later half, a thick succession of sediments was deposited in a geosyncline. The sediments were of mixed nature, mostly shale, graywacke, sandstone, chert, and rarely limestone. Deposition of the Blue Canyon formation was followed by submarine volcanic eruptions which gave rise to thick beds of volcanic materials—flows, tuffs, and agglomerates—constituting the Tightner formation. There were short breaks in the volcanic outpourings, during which argillaceous and arenaceous sediments were laid down. These sediments were rather thin and local. With the waning of volcanism, there was deposition of clastic sediments forming the Cape Horn formation. Sedimentation during Cape Horn time was frequently interrupted by a submarine volcanism giving rise to thick flows, tuffs, and agglomerates.

The frequent volcanism often left the water of the geosyncline saturated with silica which on precipitation gave rise to chert beds (Taliaferro, 1943, p. 290). There was also deposition of mixed sediments, shale, sandstone, conglomerate, and pebble beds and occasional limestone. No volcanism seems to have taken place during the deposition of the Clipper Gap formation. Sedimentation was followed by folding at some time between the close of the Carboniferous and the early part of the Triassic period, probably at the same time as the Appalachian revolution (Taliaferro, 1951). Then there was a long period of subaerial erosion between the end of the Paleozoic era and the upper Jurassic epoch.

The first record after the Clipper Gap formation is the Logtown Ridge formation comprising thick flows, tuffs and agglomerates. These indicate submarine volcanism, in a new geosyncline, during the early part of the upper Jurassic epoch. There were short breaks in the volcanic outpourings when thin argillaceous sediments were locally deposited.

The stratigraphic relation between the Logtown Ridge and Mariposa formations is obscured in the area. From reports in other areas, however, it appears that the submarine volcanism (giving rise to the Logtown Ridge formation) was followed by deposition of epiclastic sediments (constituting the Mariposa formation) with little or no disconformity between the two formations. The fact that the Mariposa formation lies directly on the Clipper Gap formation at Colfax indicates that the Logtown Ridge formation probably did not extend into this area; otherwise, this indicates a gradual northward transgression of the Jurassic sea. The presence of extensive breccia in the basal part of Mariposa formation probably signifies submarine landslides of the Clipper Gap formation in its final stage of deposition on the geosyncline. The interdigitations of conglomerate, shale, mudstone, and sandstone, are probably reflections of changing near-shore conditions and of the varied relief of the floor on which

MICROSTRUCTURE IN SLATES

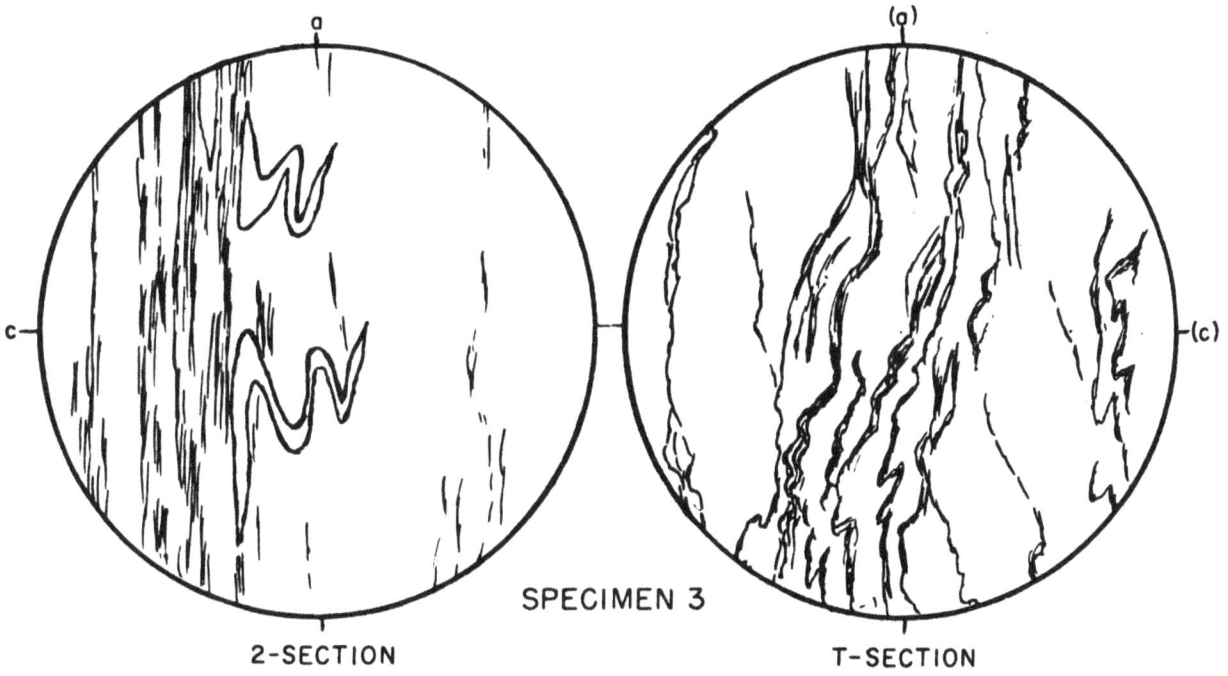

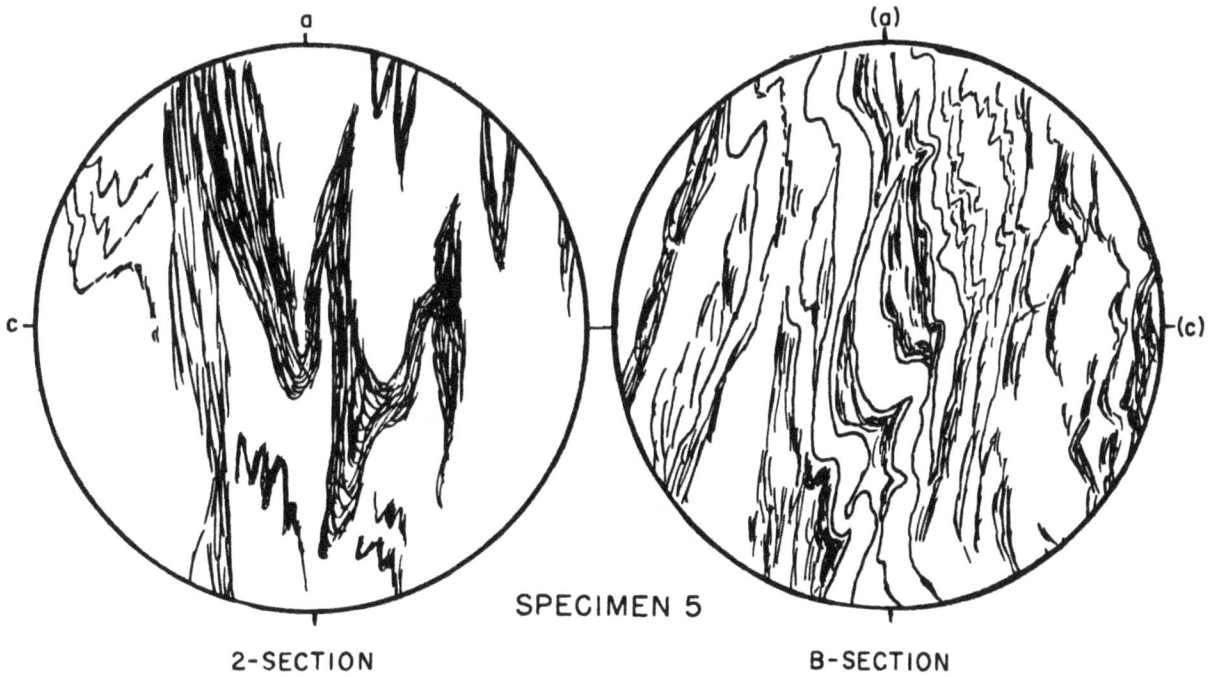

SPECIMEN 3

2-SECTION   T-SECTION

SPECIMEN 5

2-SECTION   B-SECTION

MAGNIFICATION 10X

FIGURE 6.

MICROSTRUCTURE IN SLATES

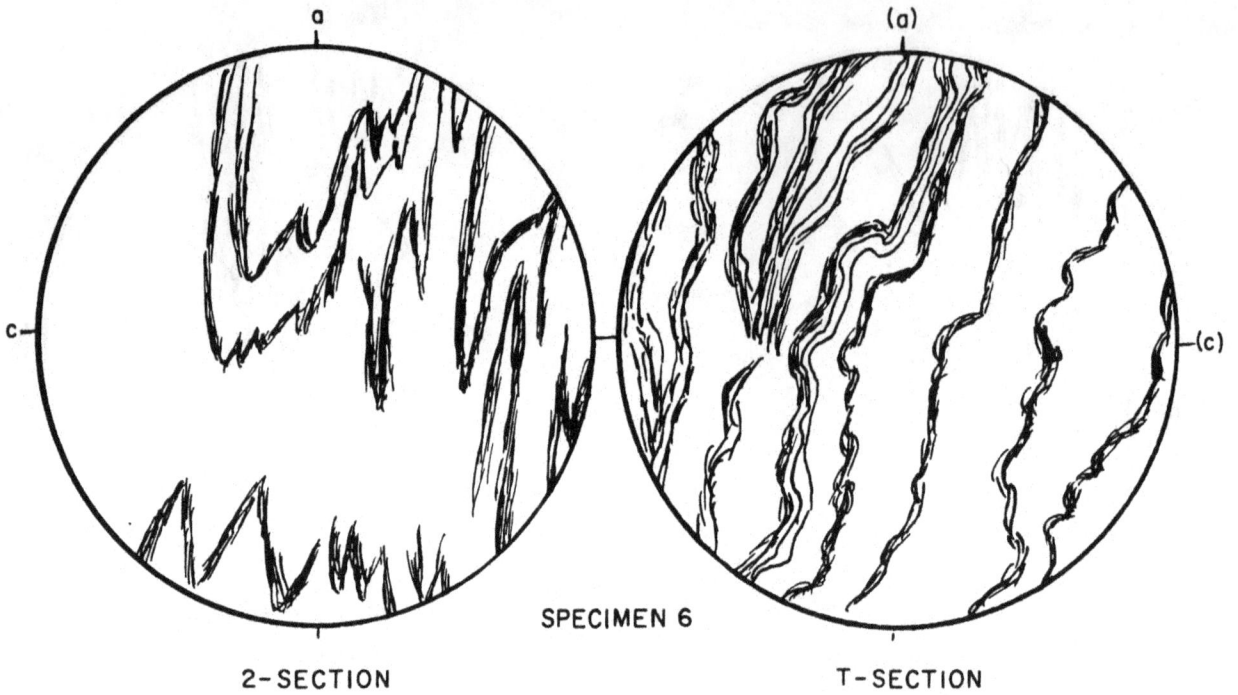

SPECIMEN 6

2-SECTION   T-SECTION

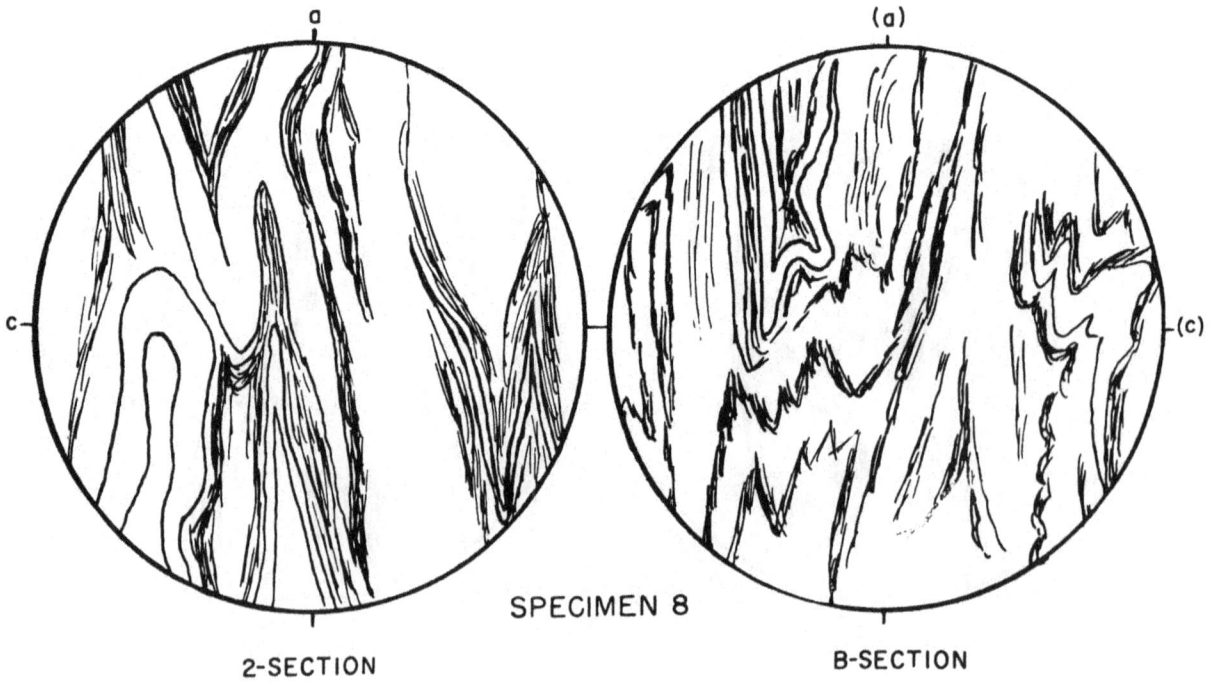

SPECIMEN 8

2-SECTION   B-SECTION

MAGNIFICATION 10X

FIGURE 7.

MICROSTRUCTURE IN METAVOLCANIC ROCKS

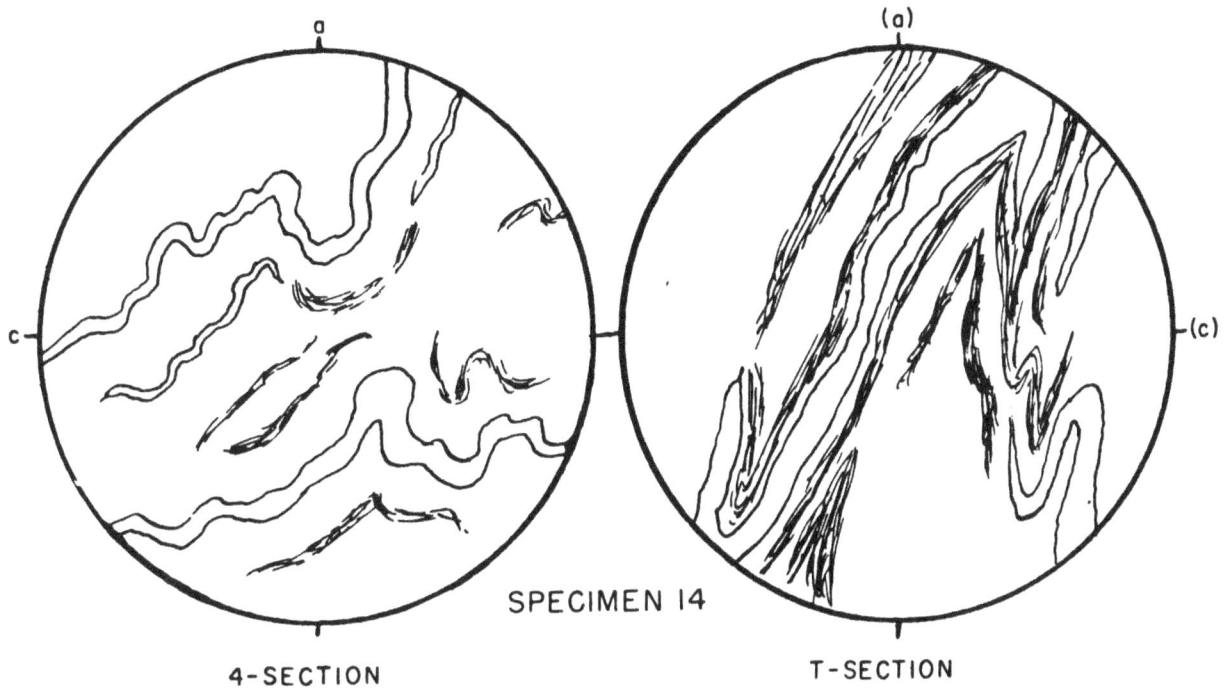

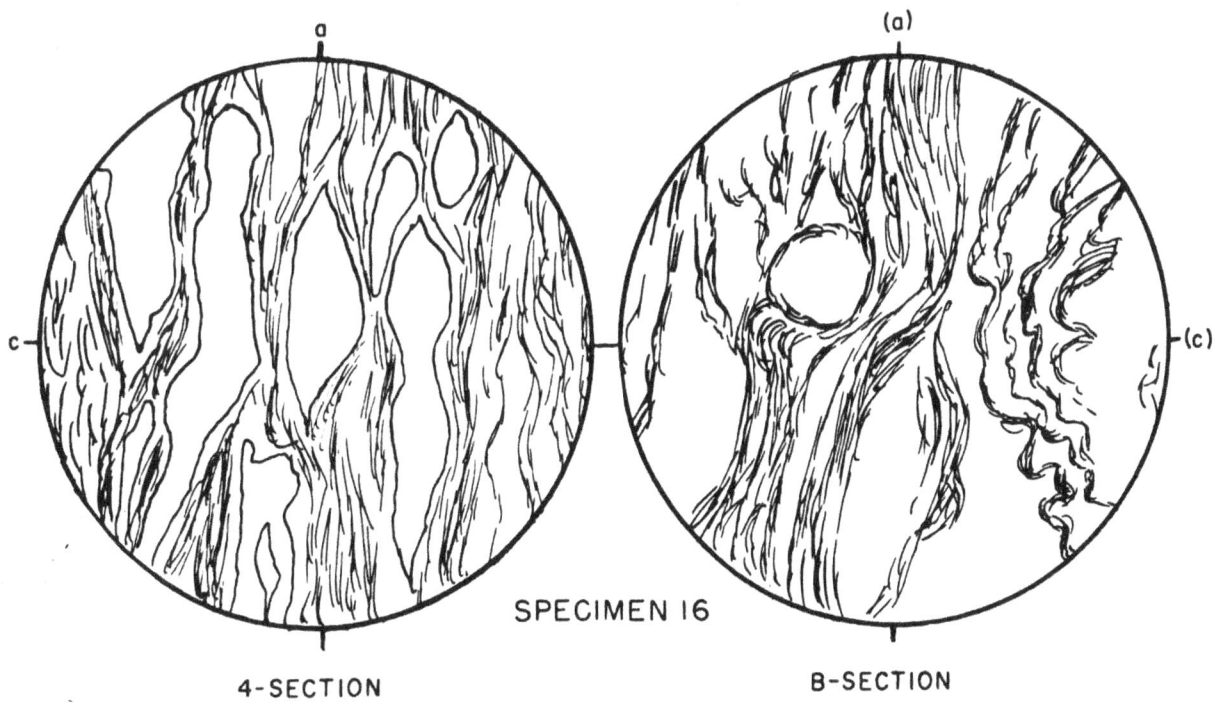

SPECIMEN 14

4-SECTION   T-SECTION

SPECIMEN 16

4-SECTION   B-SECTION

MAGNIFICATION 10X

FIGURE 8.

the sediments were laid. Such variations may also indicate pulsations in the supply. Toward the end of deposition of the Mariposa formation, but before the beginning of the Nevadan orogeny, basic intrusive bodies invaded the sediments of the geosyncline in the form of sills, dikes, and plug-like bodies.

The Nevadan orogeny was both preceded and followed by events of great importance. The slow sinking of the geosyncline was accompanied by outpourings of spilitic lava. With the onset of the orogeny, the geosynclinal sediments were compressed into a series of folds accompanied by complex faulting and thrusting; gabbro and ultrabasic rocks (mainly represented by serpentine) intruded the folded sediments. The gabbro invaded in the form of plugs and small, boss-like bodies, and granodiorite was emplaced as bosses and batholiths. The last important event in the pre-Tertiary history of the region was the introduction of quartz veins along faults and shear zones.

These events took place within a short time in the later part of the upper Jurassic epoch. The geosynclinal sediments, thus folded and faulted, gave rise to a high fold-mountain range which was exposed to weathering agencies and erosion. During the end of the Mesozoic era and early part of the Eocene period, extensive reduction of the mountain mass took place. The climate was semi-tropical—humid, one of the best environments for extensive chemical weathering. Long-continued erosion combined with chemical weathering released gold from the quartz veins. The free gold and quartz pebbles rich in gold were removed, deposited, and finally concentrated along streams, especially in the deep gravels. Because of its high specific gravity, gold became concentrated in the holes and cracks of the bedrock along the trough of the channel. Some of the finer gold, however, was caught in the flood plain sediments giving rise to rich placer deposits in sporadic spots.

From late Eocene to late Miocene, there was a series of volcanic outbursts of rhyolite in the high Sierra, causing a fall of fine rhyolitic ash which, when washed into the streams, caused damming of the streams and diversion of the drainage. The streams became overloaded with sediments and formed meanders. The volcanism was accompanied by earth movements which caused the rejuvenation of the streams. The gravels deposited along the channels during the middle Eocene were removed, carried farther downstream, and redeposited by the intervolcanic streams. These intervolcanic gravels represent an interlude in volcanic activity. With further volcanism, finer ash was deposited in lakes, streams, and lowlands, giving rise to the thick and widespread Valley Springs formation.

Following the widespread ejection of rhyolitic debris, extensive eruptions of andesite took place in the high Sierra during the Mio-Pliocene time. The latter volcanism continued up to the middle Pliocene, and was even more severe than the previous one. Volcanic ash was carried by wind and water toward the foothills; streams were again dammed. Thick masses of tuffs and volcanic bombs also fell on the land mass. When soaked with water, the ash formed a mud which started flowing down the slope, and at the same time carried pebbles, cobbles, and boulders of andesite and other older rocks that were lying loose. Mudflows not only filled the valleys but also probably spread out over the whole area. They were later stripped off, in many places, by erosion.

During the volcanic period and thereafter, earth movements caused the Sierran block to tilt westward. As this area was close to the Sierran foothills, there was little tilting within the area—1° or less; yet this small tilt was enough to cause rejuvenation of the stream.

Though the block faulting continued elsewhere in the Sierra Nevada through the Pliocene and Pleistocene periods, little record has been left in the rocks. Rejuvenation of the main stream flowing through the area resulted in formation of bench gravels. These were left at different levels and frequent intervals on either side of the streams. In a few places the deep gravels were reworked by the streams to form rich pockets.

METAMORPHISM

The oldest rocks (all of the Calaveras group) have been subjected to two regional deformations. Rocks belonging to the Logtown Ridge and Mariposa formations were subjected to only one deformation. Consequently rocks of the Mariposa formation such as shale, siltstone, and mudstone have undergone some compaction and have developed slaty cleavage. They have not advanced to the phyllitic stage. The penecontemporaneous intrusive bodies have well-preserved textures; except for some hydrothermal alteration, the minerals in these intrusive rocks are also preserved. Diabase intrusions close to serpentine bodies have invariably undergone metasomatic changes with consequent development of albite and/or prehnite.

The Logtown Ridge formation, represented by basalt flows, tuffs, and agglomerates with intercalations of slate, shows weak regional metamorphism. The tuffs and the intercalated sediments have slaty cleavage; the massive flows and agglomerates retain their texture well, the amygdules are preserved without much flattening; some of the minerals constituting the rock, however, are hydrothermally altered.

In the Calaveras group, finer-grained rocks, such as tuffs of the Cape Horn formation, belong to the greenschist facies. Except for some effects of hydrothermal activity, the coarser rocks are practically unaltered. The argillaceous sediments belonging to this formation are altered to slates and phyllites. The rocks of the next older formation—the Tightner, belong to the albite-epidote-amphibolite facies. The rocks are schistose and have strongly aligned amphiboles. Amphibolites presumably belonging to the Tightner formation and exposed at the anticlinal crests of the Clipper gap formation in the western part of the area, are often garnetiferous, and thus probably belong to the amphibolite facies. They are intensely brecciated in some places and show a later retrograde metamorphism.

In the Blue Canyon formation (the oldest member of the Calaveras group), the argillaceous sediments are locally converted to phyllites and schists, and in a few places to garnetiferous schist (evidently resulting from contact metamorphism caused by some underlying intrusive body).

The predominant type of metamorphism is regional. Rocks belonging to the Calaveras group are usually of the green-schist facies or albite-epidote-amphibolite facies and rarely of the amphibolite facies. The preservation of relict texture such as amygdules and phenocrysts in the massive igneous bodies belonging to the Calaveras group shows that the stress was of a mild nature. The heat was presumably moderate as in most cases the minerals were formed under moderate to low temperature. Locally they show retrograde metamorphism, as is evident from the presence of chlorite or stilpnomelane.

Rocks showing contact metamorphism are rather rare. Diabase and andesite intrusions have an indurating effect on argillaceous sediments immediately adjoining. Hornfels has not been noted. Shale occurring as an inclusion in a serpentine intrusion does not show contact metamorphism; this may support the hypothesis of cold intrusion of serpentine. Metasomatism plays an important part in some places. Diabases, particularly those close to serpentine intrusions, have suffered extensive soda metasomatism, that is, plagioclase (usually labradorite) has been replaced by albite. Metasomatic replacement of plagioclase by prehnite is a common instance of lime metasomatism in rocks adjacent to ultrabasic intrusions. Serpentine bodies have been subjected to hydrothermal and metasomatic changes along their margins and shear zones where talc rocks and carbonate rocks have been developed at the expense of serpentine. Silicification of argillaceous sediments in places has given rise to cherty slate.

Stress prevailing at the time of regional metamorphism was strong enough to fold the rocks tightly and cause development of schistosity. There are also rotational effects in knots in phyllites, and flattening of the lapilli in tuffs. Such effects are visible in finer-grained rocks. The schistose amphibolite, with amphiboles showing a strong orientation, may indicate the existence of moderate directional stress during metamorphism.

TERTIARY CHANNELS

Eocene Channels. Lindgren (1911) has suggested one main channel to include most of the big patches of auriferous gravels in the eastern half of the area. According to him, this channel enters the area at Michigan Bluff, passes near Sage Hill, runs west to the Paragon Mine, then north and west to the Mayflower Mine and then south to the San Francisco Mine. From here on, this channel takes a less discernible course to Yankee Jim, Smith's Point, then to King's Hill, Big Dipper Mine, Morning Star Mine, Iowa Hill, and finally north to Gold Run.

From Yankee Jim on, this channel travels somewhat upgrade in its flow to Dutch Flat. He attributed the apparent upward flow to the westward tilting of the Sierran block. In his opinion, as the stream was flowing south to north and not south-southeast to north-northwest, the west-southwestward tilting of the Sierran block caused the persent-day upgrade of the supposed course of the Ione Channel, that is, this tilting produced higher bedrock elevations at Big Dipper Mine, California Morning Star Mine, and Iowa Hill successively from the almost level course of the stream.

Regarding the course of the channel, the writer agrees with Lindgren as far as the site of the Foresthill placer mines. From the Mayflower Mine onward, however, the channel more plausibly ran south to Baltimore, thence southwest to the San Francisco Mine and Pond's Pit, and finally south to join with any stream flowing in the general direction of present flow of the Middle Fork of the American River. The elevation of the bedrock at Pond's Pit is not too great to have prevented the channel from taking that course.

A little west of this channel is another smaller channel, coming from the direction of Cottage Home Hill and extending through Buckeye Mine, Keystone Mine, Adam's Pit, Smith's Point, Small Hope Mine, Yankee Jim (where the channel forms a bar), thence to the placer diggings south of Devil's Canyon and finally towards Grey Eagle Mine.

The most important channel is the one which comes from Gold Run, enters the area at Roach Hill, thence crosses Indian Canyon to form the lucrative Iowa Hill Mine deposits, recrosses Indian Canyon to form the Iowa Hill deposits, and then passes through the California Morning Star Mine. Thus the channel follows a meandering course to the Morning Star Mine. From there, it flows east, then southwest to the Big Dipper diggings where it receives a tributary coming from Grizzly Flat. From the Big Dipper Mine, the channel flowed southwest to join with a tributary coming along King's Hill and King's Hill Point. At Roach Hill the former channel joins another tributary extending from North Succor Flat via Penn Valley and Monona Flat.

Photo 33. Canyon profile, North Fork American River (center), shows two stages in canyon cutting; view east from Colfax.

The channel along King's Hill probably drained the Second Sugar Loaf area, and flowed through King's Hill Point to join with the Roach Hill-Iowa Hill-Big Dipper Mine channel; and the joint channel might have run in the general downstream direction of the North Fork of the American River.

As the area is close to the foothills, the tilting due to the Sierran block movement was here probably less than 1°. This is evident from a study of the profile of the Eocene surface from west to east. Furthermore, a greater tilt would go against the possibility of flow of any stream from Michigan Bluff to Paragon Mine. Such an insignificant tilting would cause no appreciable difference in the relative elevation of the bedrocks of the two streams, one flowing north, the other north-northwest. The writer checked the elevation of bedrocks at the mines from the topographic maps and found that the bedrock elevation at Iowa Hill is higher than at California Morning Star Mine, while the latter is higher than the bedrock elevation at Big Dipper Mine. This suggests that the flow direction of the Eocene channel was from Iowa Hill to California Morning Star Mine and thence to Big Dipper Mine.

Lindgren did not include Roach Hill in this channel for the obvious reason that the elevation of the bedrock at Roach Hill was 300 feet higher than at Iowa Hill. The bedrock elevation at Gold Run is about 350 feet higher than at Iowa Hill. This also suggests that the flow direction of the Eocene channel was from Gold Run to Roach Hill and thence to Iowa Hill.

Intervolcanic Channels. In most of the placer mines, "auriferous gravels" are represented by deep gravels. Bench gravels and other flood plain deposits belonging to the same formation are generally lacking except in a few places. In one such place, Elizabeth Hill Mine, bench gravels, sands, and carbonaceous shale are present. These are overlain by the Mehrten formation without any intervening Valley Springs formation, indicating no intervolcanic stream eroded the flood-plain deposits left by the Eocene stream. In other placer mines, the deep gravels are generally overlain by gravels of the Valley Springs formation, believed derived from the auriferous gravels and redeposited by the intervolcanic streams along with some rhyolite detritus. On this basis, this writer finds that for most of the Eocene channels there is a corresponding intervolcanic channel following almost the same course as the Eocene stream.

According to Lindgren, the Ione channel from Foresthill onward was different from the intervolcanic channel which, in his opinion, agreed with present-day drainage. This writer suggests that the intervolcanic streams followed more or less the same courses as the Eocene streams, and that the main channels of both periods followed the general direction of present-day main drainage. It should be noted, however, that the drainage patterns of Eocene and intervolcanic channels were different from that of present-day channels, that is, the Eocene and intervolcanic streams, overloaded with sediments and flowing on a flat surface, formed meandering courses, whereas the present-day streams, being in the youthful stage, flow more or less straight.

ECONOMIC GEOLOGY
Gold Placer Mines

When the easily-won placer gold deposits along the banks and beds of the present-day rivers and streams were exhausted, search for pay gravels continued higher up in the hills. Thus surface gravels such as those at Iowa Hill, Yankee Jims, were explored. This led to the search for the continuation of such channels below the thick cap of volcanic muds and boulders. In most places, two or three channels are superimposed, the older channel being generally richer in gold content than the younger ones. Thus the Eocene channels contain more placer gold than the intervolcanic ones which, moreover, derived most of their placer gold by reworking the gravels of the former channels. The major part of the placer gold found along the banks and beds of the present channels, was derived principally from the reworking of the auriferous gravels of the Eocene and intervolcanic channels.

Before hydraulic mining was introduced the placer deposits in the area were worked by drifting. When restrictions were imposed on the hydraulicking, drift mining once more became an important method of mining. In the placer deposits, the pay-streaks were usually on or near bedrock; and if there was a thick unproductive andesitic rhyolitic mantle, mining by hydraulicking became uneconomical, especially when there was insuffi-

Table 3. Summary of structural data.

Age of deformation	Deformation		Formation affected	Large-scale structure	Small-scale structure	
					Drag folds	Lineation
UPPER JURASSIC	NEVADAN OROGENY (Second deformation)	Later phase	Mariposa	Moderately plunging folds; locally steeply	not known	not known
			Cape Horn (Calaveras)	nil	steeply plunging folds	steep lineation
		Earlier phase	Mariposa	Horizontal or gently plunging folds	not known	not known
			Cape Horn (Calaveras)	Horizontal or gently plunging folds (tight isoclinical folds)	Horizontal or gently plunging folds (tight isoclinal)	nil
Some time between Carboniferous and Jurassic periods. Possibly at the end of Palezoic.	First deformation		Cape Horn (Calaveras)	Horizontal or gently plunging folds	Horizontal or gently plunging folds	nil

Photo 34. Middle Fork, American River, showing "mountain-valley stage" (right central part of photo) and "canyon stage" (center foreground). View west-southwest from Horseshoe Bar.

cient water or fall of water. In such cases, drift mining was the only feasible method. In several mines, however, where the mantle over the pay dirt was rich enough to pay for washing, hydraulicking successfully replaced drifting.

The drift mines in the area were worked by adits driven along or close to the trough of the channel, so that the pay dirt could be worked directly by the levels or raises. When there was a thick overburden, shafts, usually vertical, but occasionally inclined, were sunk to or beyond the drift which ran along the course of the

old channel. In order to prospect the whole width of the channel, crosscuts at right angles to the drift were driven on both sides. In a few mines, adits were driven in the competent bedrocks, just a few feet below the old channel bed, and the pay dirt lying in the channel bed was tapped by use of raises. By this method, mining was done without much timbering. The channel gravels in some mines, however, were cemented and were strong enough to stand without timbering. The drifts were driven in such a way as to allow drainage out of the entrance adit. Besides the main shaft for hoisting ores

Table 4. Gold placer mines.

Claim, mine, or group	Sec.	T.	R.	B & M	Remarks
Adams Pit	N½ 27	14N	10E	MD	On divide between First and Second Brushy Canyons. North part worked by hydraulicking.
Big Dipper	3 & 4	14N	10E	MD	Worked by hydraulicking 1858-1882, then by drifting until connection made with Calif. Morning Star Mine. Gravel said to have averaged $6.00 per ton.
Big Spring	33 & 34 / 3	14N / 13N	10E / 10E	MD / MD	On west side of Dardanelles. mine. Opened in 1854 as drift mine and worked until late 1800's. Part of work done through 1400' Big Spring adit, and part through adit shared with Dardanelles. In 1882 reported to have had production of $150,000. Pay gravel 5' thick, well cemented. Much of property now subdivided into town lots.
Buckeye	11 & 14	14N	10E	MD	Drift mine. Adit southeast into hillside. Mining confined to boulder bed resting on bedrock.
Campbell	SE¼ 4	14N	10E	MD	Two adits run N.80°W. and N.40°W. in bedrock. Remains of milling plant, concentrator, sluice boxes.
California Morning Star	33 & 34	15N	10E	MD	Drift mine. Portal and mill east side of Indian Creek. Drift driven eastward into bedrock below channel, gravels worked through raises. Connected with drift from Big Dipper. Average width of channel 200-300'. Pay gravel extended several inches into slaty bedrock. Part of placer deposit worked by hydraulicking. Average yield said to be $7.00/ton.
Copper Bottom	25	15N	10E	MD	Upper run or bench deposit about 50' above main channel worked in 1922. Deeper channel reached through an adit 400' and a slope 300', requiring a lift of 120' for water.
Dardanelles (Oro & Dardanelles)	34 & 35 / 2	14N / 13N	10E / 10E	MD / MD	One of major gold sources in Foresthill area. Worked almost continuously by drifting and hydraulicking from 1853 to 1890. By 1890 credited with total putput of $2,000,000. Intermittently worked by drifting from 1890 to 1930. Main source of production was well-cemented, 5' lower gravel overlain by coarse gray sand. Mined to average width of 75'. Property now site of sawmill and landing strip of Stockton Box Co.
Elizabeth Hill	SE¼ 5	14N	10E	MD	Drift mine NE part of King's Hill, later worked by hydraulicking. Tunnels and shafts probably used to supply water from nearby creek or wells. Said to be exceptionally rich in coarse gold.
Gleeson Diggings	S½ 4 / N½ 9	14N / 14N	10E / 10E	MD / MD	Placer deposit, extensively hydraulicked. Several adits at eastern margin: N.75°W.; N.55°W.; S.E.; all caved. Water supply for hydraulicking presumably from Refuge Canyon through tunnels in bedrock.
Golden Streak	35	15N	10E	MD	Drift mine. Shaft driven in mudflow capping gravel. Closed in 1951.
Goodwin	S½ 3	14N	10E	MD	Drift mine on western flank of Prospect Hill. Two adits, one N.75°W. Both caved. Diggings south and northwest of portals. Large hydraulic pit. Trenches at regular intervals indicate systematic sampling.
Grey Eagle	6	13N	10E	MD	Portal of main adit runs south in bedrock on south side of Owl Creek. Channel near Spring Garden extensively prospected 1915-16. Considerable amounts of cemented gravel run through old stamp mill, but average returns evidently not satisfactory.
Haymes	8	14N	10E	MD	Drift mine. Two adits driven, the older trending north; other N.40°W. along Ione gravels resting on bedrock. Problem to keep roof from caving.
Hazard & Weske	NE½ 20	14N	11E	MD	Worked for about 3000', from Weske mine. Incline was run downstream several hundred feet into Muir Tunnel claim.
H&H (Hendrickson & Hoover)	18 / S½ 7	14N / 14N	10E / 10E	MD / MD	Drift mine on NW flank of King's Hill; closed in 1948. Two adits and some open pits. Eastern adit runs N.15°W. about 720'. Filled with water in 1951. Other adit blocked with boulders. In east drift, gravel bed rests on bedrock, underlies mudflow, is 1½-2' thick, consists of slate and metavolcanic rock. Rounded white quartz gravels of Ione type rare.
Independent, New Jersey, and Jenny Lind	W½ 36 / E½ 35	14N / 14N	10E / 10E	MD / MD	Claims under town of Foresthill. Primary stream deposit about one mile in length, flowing west, between canyon slope and deep blue lead. Expectation of finding western segment of channel be point where it presumably had been cut off by deep blue lead inspired work after pioneer miners.

Table 4. *Gold placer mines--Continued*

Claim, mine, or group	Sec.	T.	R.	B & M	Remarks
Iowa Hill	27, 33, 34	15N	10E	MD	Originally a drift mine, worked by hydraulicking before 1880, along Indian Canyon front. Main adit runs SE. Worked intermittently until 1935 by Iowa Hill Gold Mining Co. Reopened in 1940 and 1944. Probably yielded more than $10,000,000 between 1849 and 1911.
Jupiter	2	14N	10E	MD	Drift mine, portal southeast of Grizzly Flat. First part of drift trends NE in bedrock, then follows trough of channel (about 4500'). Crosscuts on both sides. Gravel bed 1-1½' thick contains rounded/subrounded pebbles of quartz and chert—underlies mudflow and overlies bedrock. Some places andesite mudflow rests on bedrock, other places bedrock boulders underlie mudflow. Worked 1937-42, by a Mr. Carlson. Except for a few rich pockets, gold reported sparsely distributed in pebble bed overlying bedrock.
Keystone	N½ 22 S½ 15	14N 14N	10E 10E	MD MD	Hydraulic pit. Rhyolitic sands and gravels overlie basal boulder bed of probable Eocene age.
Little Indian Creek	26, 27	15N	10E	MD	Drift mine. Western adit runs N.60°E. into SW side of Strawberry Flat. Drift driven along bed of rhyolitic sand and gravel overlying bedrock and underlying mudflow. (Intervolcanic channel?) Upstream adit driven from creek bed N.60°W. into bedrock. Water-filled.
Mayflower	23,24	14N	10E	MD	Mine first worked by hydraulicking, then drifting. Same lead as bottom lead in Paragon mine. Gravel 2-14' thick yielded average of $7.00 ton. Cemented and needed milling. Upper lead, 150' above main channel said to have produced much gold. In hydraulic pit no adit can be seen, no bedrock exposed.
Michigan Bluff	21	14N	11E	MD	In western ends of mines auriferous gravels are about 40' thick, reported profitable. Total yield 1849-1911 about $5,000,000.
Occidental	3, 34	14N 15N	10E 10E	MD	Drift mine owned by a Mr. Carlson. Inclined shaft sunk NNW, 2 levels 60' apart. Lower level follows trough of channel with boulders of bedrock. Coarse gold reported in pockets. Upper level driven in thick gravel bed of rounded/subrounded pebbles, cobbles, boulders—mostly quartz-intercalated with rhyolitic sands containing abundant biotite. Extensive gravel beds on both sides of the level appear to be flood-plain deposit. Paystreak in ferruginous conglomerate associated with ferruginous shale parting. Gold recovered from upper level reported finer than that from lower. Mine has small milling plant.
Old Jupiter	11	14N	10E		Drift mine on west flank Long Point Hill. Adit running S.30°E. has caved at portal.
Paragon	19, 30, 24	14N 14N 14	11E 11E 10E		Hydraulic mine. Produced $2,651,000 prior to 1902. Pay reported to be mainly in lower 150' (between upper Paragon lead and bedrock). Lower and upper leads worked by drifting and then by hydraulicking (Mr. William Wilson) during winter when adequate water available from Volcano Canyon. Said to have been profitably worked recently. Large amount of fine gold reported lost with heavy minerals in sluice boxes.
Penn Valley	NW¼26	15N	10E	MD	Drift mine on south side of Indian Creek. Adit runs through hard metavolcanic rock so timbering unecessary. 1500' drift runs SE. Two short raises to metavolcanic and serpentine gravel and boulder beds. Quartz gravels nearly absent. Mine intermittently worked by a Mr. Henderson.
21 Placer	S½3 N½10	14N	10E	MD	Old drifts not worked for 50 years. Part of placer deposit has been mined by open-cut method.
Pond Pit	3 & 4	13N	10E	MD	In Todds Valley. Extensively worked before 1875 and proved to be a large producer. Deposit could not be worked further because of increasing overburden. Drifting not feasible because of low gold content.
Randall	25	15N	10E	MD	Drift mine. Old adit 600' long reported to run SE into hill; new one N.30°W. Mining confined to thin rhyolitic gravel bed overlying bedrock and underlying mudflows; apparently on a small intervolcanic channel.
Roach Hill	27, 28	15N	10E	MD	Drift and hydraulic mine. Bedrock exposed between tailings. One drift on west flank of hill worked intermittently on small scale.
San Francisco	33 & 34 3	14N 13N	10E 10E	MD MD	Drift and hydraulic mine between Pond mine on west and Big Spring mine on east. Chief source of gold was well-cemented quartzitic gravel on slate bedrock.

Table 4. Gold placer mines—Continued

| Claim, mine, or group | Location | | | | Remarks |
	Sec.	T.	R.	B & M	
Small Hope	27	14N	10E	MD	Drift and hydraulic mine on north side of First Brushy Canyon. Adits driven to prospect deposit of 30-40' gravels, sands, shale overlain by andesite mudflow. Channel also contained cemented gravels of intervolcanic period.
Star United	27	15N	10E	MD	Old mine fronting on Indian Canyon 1½ mi. E of Iowa Hill Mine. Extensively mined by drifting and hydraulicking.
Strawberry	26, 27	15N	10E	MD	Drift mine under lease to M. Capurro in 1950. Portal on S. bank of Indian Creek. Drift runs SE about 1200'. Follows trough of small channel. Mining confined to nonpersistent rhyolitic gravel bed overlying bedrock and underlying mudflow. Channel evidently intervolcanic.
First Sugar Loaf	32, 33	15N	10E	MD	Eastern half extensively worked by drifting and hydraulicking. Western half not hydraulicked owing to thick overburden of mudflow andesite.
Volcano	19	14N	11E	MD	Adit runs north in gravel bed overlying serpentine bedrock and underlying mudflow.
Welcome	7	14N	10E	MD	Drift mine on south flank of King's Hill. Westernmost adit trends N.15°W. Eastern adit trends N.20°W. for 700' but is caved. New adit being driven parallel to latter in harder bedrock in August 1951. Mine equipped with big water pump, small mill, and concentrator.

Photo 35. North Fork American River, view southwest from Big Bend. Upper part of canyon profile (right) shows "mountain-valley stage"; lower part of profile, "canyon stage".

Photo 36. V-shaped gorge of North Fork, American River, indicating "canyon stage". View southwest.

or pumping water, there were one or more vertical shafts for ventilation in mines, especially where the drifts extended for long distances.

The major placer deposits in the area were at one time or another mined by hydraulicking. In most cases water had to be brought in open ditches, pipes or flumes from long distances. These ditches and flumes were high enough above the mines to furnish a satisfactory pressure for the giants. In rare cases water was pumped from neighboring creeks. The majority of placer mines could be operated from December to May. With the beginning of the winter rains and the later melting of snows, the creeks had increased volume of water, and adequate supply through ditches and flumes could be obtained.

The gravels varied from soft, easily washed material to cemented gravels that had to be loosened by blasting or by jets of high-pressure water. If the gravels had a

Table 5. Gold lode mines.

Claim, mine, or group	Location				Remarks
	Sec.	T.	R.	B & M	
American Bar Quartz	33	14N	11E	MD	On slope north side Middle Fork of American River. Worked through adits at different levels. Several quartz veins 1-3' thick emplaced in slate which is also cut by basic dikes and sills carrying much pyrite. Worked intermittently since 1895. Ore yielded more than ½ oz. gold per ton.
Annie Laurie	24 25	14N	9E	MD	North side of Foresthill road where crossed by Gillis Hill thrust. Vein strikes north and dips steeply east. Worked through a few adits and an open cut. One adit trends north and one northwest.
Big Oak Tree Quartz	S½ 33	15N	9E	MD	Vein along contact zone of diabase and gabbro. Reported average width 18", strike N.80°E. and dip nearly vertical. Surface trace few 100' south of Rising Sun vein. Two shafts, deeper one 175' on 65° incline with 3 levels. On west, vein turns northwest where there are 3 more shafts. One shaft on eastern part of vein had new overhead frame. Ore reported to average $28.00 a ton.
Bauer	24	14N	9E	MD	North side of Foresthill road. Two adits trend north-northwest, one more than 300', other 80' driven to prospect ledge with reported yield of $3.50-$5.85 a ton across a width of 20'.
Black Oak	SW¼ 35	14N	9E	MD	Several adits driven to prospect gold-bearing vein, one trending west, one northwest, third southwest. Vein reported to strike N.10°W. and dip 52° east. Vein swells and pinches vertically and horizontally. Southern (uppermost) adit said to follow vein 528' farther south. Just outside portal of adit is 23' shaft from which 120' drift runs.
Brunn	NW¼ 36	15N	10E	MD	Nonpersistent quartz vein on west bank of North Shirttail Canyon at junction with Snail Canyon. Less than 1' thick. Trends N.40°W. and dips vertically. Emplaced along shear plane. Extends southeast across creek. From shaft sunk on west bank of North Shirttail Canyon, drift was driven below canyon bed to east side. Lode consisted of stringers of quartz a few inches thick, but very rich. Tunnel below canyon caved because of water seepage.
Brushy Creek	26 & 35	14N	9E	MD	Several adits and pits in quartz veins and stringers. Pits shallow, drifts short. Last adit downstream trends N.50°E. Vein about 2' wide cuts into sandstone and mudstone, dips 40° SSW.
Doherty	30 & 31	14N	11E	MD	Gold lode in intimate association with ankerite vein 1'-5' wide. Country rock serpentine. Owner claimed ore yielded assay values of $53.00 per ton.
Doherty Prospect	30	14N	11E	MD	North of Doherty mine and associated with same ankerite vein. Mineralized quartz stringers in serpentine country rock. Owner claimed ore yielded assay values of $18.00-$19.00 per ton. Gold occurs free and is rather fine-grained.
Drummond	1	14N	10E	MD	On west flank of Cottage Home hill. Vein strikes N.60°W. and dips steeply NE. Width of vein varies. Lower adit running N.25°E. still open. Small milling and washing plant.
Ferrier	SE¼ 30 SW¼ 29	14N	11E	MD	In Ladys Canyon near Mosquito Ridge road. Diggings in quartz stringers in slate and metavolcanic rocks at contact of serpentine body. Andrew Ferrier, discoverer, said to have taken a small fortune from the claim.
Garbe & de Maria	SE¼ 30 SW¼ 29	14N	11E	MD	In Ladys Canyon south in Ferrier diggings. Reported to have contained occasional pockets of coarse crystallized gold lying in flat seam. Mined through network of crooked drifts.
Hinchy	25	14N	9E	MD	Top of Big John Hill. Circular shaft. Country rock Mariposa conglomerate.
Homestake	6	13N	11E	MD	On divide between Blind Canyon and Volcano Canyon. Adit runs northwest in serpentine. Now covered by debris.
Kings Hill Quartz	SE¼ 7 SW¼ 8	14N	10E	MD	Vertical shaft, caved at top, connected with two inclined shafts dipping 35°NE. Quartz vein 2' thick pinches into stringers along a shear zone.
Kings Hill Quartz ... ct	8	14N	10E	MD	Several quartz stringers and veins prospected by shallow pits and trenches on south flank of King's Hill.

Table 5. *Gold lode mines—Continued*

| Claim, mine, or group | Location | | | | Remarks |
	Sec.	T.	R.	B & M	
Live Oak Ravine.............	23	14N	9E	MD	Four adits and two pits along south bank of Live Oak Ravine. Easternmost tunnel (closest to ravine) locked. Trends N.10°W. Adit next to it trends same at slightly higher elevation. Third adit trends N.30°E., almost at same level as second. Two open shallow pits west of third adit. Fourth adit runs N.5°W. south of pits in conglomerate band along tributary of Live Oak Ravine. Reported that "the lessees . . . were prospecting in 1935-36 . . . in a winze . . . giving good assays from a stringer."
Providencia (Providence).....	25	15N	10E	MD	On steep west bank of North Fork of Shirttail Canyon upstream from junction with Snail Canyon. Vein strikes N.60°W. and dips 70°NE., about 3′ wide. Worked through adit along vein and by winzes, particularly along footwall. Some good assays reported from vein. Remains of old stamp mill about 100 yards downstream from junction of Snail Canyon and North Shirttail Canyon.
Rising Sun.............	33	15N	9E	MD	About a mile northwest of Colfax. Vein strikes NE and dips 85°SE. Country rock is saussuritized gabbro, locally brecciated and cut by quartz stringers. Vein said to have averaged 18″ in width and to have swelled to 4′ in thickness. Pay chimney, rich in free gold, reported at depth, bounded by cross fault on one side. Three shafts and several shallow pits. Mine closed about 1935.
Roach Hill Quartz............	28	15N	10E	MD	Quartz ledge adjacent to small serpentine body on northwest flank of Roach Hill. Ledge about 2′ wide, strikes NNE and dips 50°ESE. Inclined shaft along dip of vein, and at least two levels (lower level under water in 1951).
Specimen Gulch..............	W½ 25 E½ 26	14N	9E	MD	Several quartz veins and stringers along Specimen Gulch prospected with shallow pits, trenches, and short adits. Last Chance prospect (next-to-last downstream) has two short drifts, one trending N.80°E.; other N.10°E. Stringers of quartz, a few inches thick along minor faults with some gouge. Country rock mudstone and graywacke. Divide prospect at next draw on south side of gulch. Adit southwest of Last Chance runs N.40°W. Fine-grained gold reported.
Three Queens (Four Aces).....	31	14N	11E	MD	In Volcano Canyon. Lode confined to contact zone of serpentine body. Adit on east bank runs east. Portal blocked. Adit on west bank partly caved. Quartz vein 3-4′ wide strikes NW and dips 30°-40° NE. High-grade ore in small bunches evidently formed along hanging wall by repeated offsetting along vein. Near southeast end of adit, thin, gold-bearing veinlet strikes N.20°W. and is nearly flat.
Old Vore (Chubb)............	3	13N	9E	MD	Two miles east of Applegate on east side of Codfish Creek. Several adits driven into vein at different elevations. Three trend approximately north, another due west. A few other adits and pits north of mine. Country rock slate, quartzite, (cherty), or metasomatized volcanic breccia of Clipper Gap intruded by massive quartz vein and several quartz stringers. Main lode close to local fault zone and thin serpentine body. Vein thickens and thins from a few inches to 6′. Fairly wide on surface where it is almost barren. Bottommost tunnel, where quartz vein breaks up into stringers, being worked in 1952. Mineralization usually confined to country rock forming walls of vein. Flat-lying quartz veins or stringers along joints of country rock rather barren. Mine worked on small scale from time to time. Hydraulicking done along hillsides around 1900.
West Quartz............	31	14N	11E	MD	Adit runs east-southeast along strike of lode. Portal just below Mosquito Ridge road. Quartz vein breaks up into stringers inside tunnel. Some surface workings above road.
Whiskey Tunnel.............	W½ 30	14N	10E	MD	On west bank of North Fork of American River. Adit runs northwest into hillside and follows shear zone in slate and metavolcanic rocks of Cape Horn. Lode breaks up into thin stringers. Mineralization confined to wall rocks. Two cross cuts beside main drift. Tunnel used for growing mushrooms in 1951.

Table 6. Prospects.

Location				Remarks
Sec.	T.	R.	B & M	
17	14N	9E	MD	Vertical shaft 40-50'. Under water in 1950. Country rock is conglomerate with bands of slate and cherty quartzite. Mineralized quartz vein contains abundant sulphides such as pyrite, arsenopyrite, and tetrahedrite. Owner reports 35¢ in gold per ton by chemical analysis, but claims to have recovered $150.00 a ton by roasting the ores first.
W½ 29	14N	11E	MD	Prospect in serpentine body, at junction of two roads. Vertical shaft full of water.
S½ 36	14N	10E	MD	On divide between Snyder and Blind Canyons at sharp bend of road to Three Queens mine. Short adit into quartz vein.
36	14N	10E	MD	On Mosquito Ridge road near B.M. 2912. At contact of serpentine and metavolcanic country rock.
1	13N	10E	MD	Diggings along north bank of Middle Fork of American River. Quartz stringers in Cape Horn.
N½ 15	14N	10E	MD	Flat-lying quartz vein exposed along north bank of Shirttail Canyon. Vein, of variable thickness, strikes N.30°E. Several prospecting pits and adits along vein.
S½ 1 ·	13N	9E	MD	Mineralized quartz stringers between Logtown Ridge and Cape Horn. Several prospects and a west-trending adit along Ponderosa Way, where road crosses thrust zone.

Table 7. Chromite deposits.

Claim, mine, or group	Location				Remarks
	Sec.	T.	R.	B & M	
Washout....................	N½ 30	14N	11E	MD	East of Paragon placer deposit, on northwest side of Volcano Canyon. Being developed in 1952. Fairly good road to Paragon Mine, from there road ungraded, very narrow. George Ford and John Lower reported to have operated deposit in 1941-42, producing 450-500 tons. More recently, tunnel driven due west to join vertical manhole. Most time devoted to development work. Up to 1952 only 10-15 tons produced. Chromite occurs as massive lenses in serpentine. Exposed along tunnel and manhole and appears to contain considerable quantities of ore. Reported to have yielded assay of 49% Cr_2O_3, but iron content of 3% was higher than acceptable to buyers.

Table 8. Asbestos deposits.

Location				Remarks
Sec.	T.	R.	B & M	
SE¼ 35 NE¼ 2	15N 14N	9E 9E	MD	Large number of shallow pits in Burnt Flat north of Iowa Hill road. Country rock is serpentine. Asbestos of amphibole variety.
E½ 28	15N	10E	MD	In northwest corner of Roach Hill. Small serpentine body contains pockets of asbestos of tremolite variety. Not of spinning quality.
S½ 33	15N	10E	MD	South of California Morning Star Mine. Serpentine contains sparsely distributed pockets of asbestos. Several pockets worked in 1950 by a Mr. Gary.
33	15N	·.·	MD	Serpentine body north of California Morning Star Mine. Exposed along road to Iowa Hill Mine. Contains some asbestos-bearing rock. Extensive bulldozing has exposed bedrock.

clayey matrix, the washing was done better by hydraulicking. The quantity of gravels excavated depended on the head used on the giants, but high pressure water was more likely to carry the finer gold out of the sluice boxes.

The Paragon Mine is the only one in the area that was being worked by hydraulicking at the time of this investigation. The work season is from December to May when adequate water supply is obtained from the Volcano Canyon through ditches and flumes.

Gold Lode Mines

The area abounds in quartz veins, some gold bearing and profitably worked in the past. The active mines were closed down as a war measure during World War II; none have been reopened. Some prospects have been intermittently worked in recent years by men as between-jobs work.

Of the systems of fissures occupied by the veins, the most important is the one in which the vein strikes a few degrees west of north with a rather steep eastward dip.

Most of the mines are idle. Information from the State Mineralogist's Reports has been heavily drawn upon and supplemented by this writer's observations in the field.

Chromite Deposits

Pocket deposits of chromite are present in dunite or peridotite which is partly or completely serpentinized. The chromite usually forms narrow streaks, called *schlieren*. It also forms massive ore—a dense aggregate of crystals—and it may also occur as disseminated ore in which the chromite is sparsely distributed throughout country rock.

The first two types of ore, that is, the lensitic and massive types, were mined during the World Wars. The disseminated type, which is fairly abundant in the serpentine areas, has not been exploited (it has to be concentrated by milling, an expensive operation, before it can be marketed).

All the chromite mines and prospects in the area have been described in detail by Rynearson (1953). The Washout Mine was being developed during October, 1952.

Pyrite Deposits

In a few places close to the Mosquito Ridge road (west-central part of Sec. 33, T. 14 N., R. 11 E.), crystals and thin lenses of pryite are abundantly distributed in the slate and graywacke of the Blue Canyon formation. These lenses are 1/10 or 1/5 of an inch thick. In a few other places, the country rocks are locally impregnated with pyrite crystals. Such occurrences are sporadic and probably cannot be profitably worked under present conditions.

Asbestos Deposits

Pocket deposits of asbestos are sporadically distributed in the smaller serpentine bodies. Numerous shallow trenches and bulldozed areas indicate that these deposits were prospected.

The asbestos found in the area is of the amphibole variety. The fibres are parallel to the direction of the veins.

Soapstone Deposits

Soapstone is invariably present along the margins of serpentine bodies or along shear zones. It is composed mainly of talc formed by the hydrothermal alteration of serpentine. It is not known if any of the soapstone deposits in the area were ever worked for industrial purposes.

Limestone Deposits

The only limestone deposit of notable size crops out along the southeast bank of the Bear River in the northwest corner of Sec. 4, T. 14 N., R. 9 E., M.D., two miles west of Colfax. It is about 400 feet long and 200 feet thick at its thickest point, strikes north and dips steeply east. The rock is a dark-gray crystalline limestone veined with white and black calcite, takes a good polish, and has been used for interior wall facings in the old U. S. Mint and Bank of California buildings. According to Logan (1947, p. 263) the deposit was discovered in 1866. Two kilns were built and lime was burned at the deposit at some time later than the time the marble was produced. There has been no mining activity at the quarry for more than 40 years. Logan (1947, p. 263) states that the quarry face is 400 feet long.

REFERENCES

Allen, Victor T., 1929, The Ione formation of California: Univ. California, Dept. Geol. Sci. Bull., vol. 18, no. 14, pp. 347-448.

Averill, Charles Volney, 1943, Current notes on activity in the strategic minerals, Sacramento field district; California Div. Mines, Rept. 39, pp. 75-76.

Averill, Charles Volney, 1946, Placer mining for gold in California, California Div. Mines Bull. 135.

Browne, Ross E., 1890, The ancient river beds of the Foresthill Divide: California Min. Bur. Rept. 10, pp. 435-465.

California Division of Mines, 1948, The Mother Lode Country; Geologic guidebook along Highway 49, Bull. 141.

Chaney, R. W., 1932, Fossil plants found in the auriferous gravels of the Sierra Nevada: California Div. Mines Rept. 28, pp. 299-301.

Cloos, Ernst, 1935, Mother Lode and Sierra Nevada batholith: Jour. Geology, vol. 43, pp. 225-249.

Cloos, Ernst, 1947, Oolite deformation in the South Mountain fold, Maryland: Geol. Soc. America Bull., vol. 58, pp. 843-917.

Condit, Carlton, 1944, The Remington Hill flora: Carnegie Institute of Washington, Pub. 553, pp. 21-55.

Crowell, J. C. and Winterer, E. L., 1953, Pebbly mudstones and tillites: Geol. Soc. America Bull., vol. 64, p. 1502.

Curtis, G. H., 1951, Geology of Topaz Lake quadrangle: Unpublished Ph.D. thesis, Univ. California, Berkeley.

Daly, R. A., 1933, Igneous rocks and the depths of the earth. McGraw-Hill Book Co. Inc., New York, N.Y.

Eardley, A. J., 1951, Structural geology of North America, Harper & Brothers, New York, p. 249.

Fairbairn, H. W., 1935, Notes on the mechanics of rock foliation: Jour. Geology, vol. 43, pp. 591-593.

Fenner, C. N., 1948, Incandescent tuff flows in southern Peru: Geol. Soc. America Bull., vol. 59, pp. 879-893.

Ferguson, H. G. and Gannett, R. W., 1932, Gold-quartz veins of the Alleghany district, California: U.S. Geol. Survey Prof. Paper 172, 139 pp.

Haley, C. S., 1923, Gold placers of California: California Div. Mines, Bull. 92.

Hammond, John Hays, 1890, The auriferous gravels of California: California Min. Bur. Rept. 9, pp. 105-138.

Hess, H. H., 1933, Hydrothermal metamorphism of an ultrabasic intrusive at Schuyler, Virginia: American Jour. Sci., 5th ser., vol. 26, pp. 377-408.

Knopf, A., 1929, Mother Lode system of California: U.S. Geol. Survey Prof. Paper 157.

Kuenen, Ph. H. and Menard, Henry W., 1952, Turbidity currents, graded and non-graded deposits: Jour. Sed. Petrology, vol. 22, no. 2, pp. 83-96.

Lindgren, Waldemar, 1900, Colfax folio, California: U.S. Geol. Survey, Geologic Atlas, Folio 66, 12 pp.

Lindgren, Waldemar, 1911, The Tertiary gravels of the Sierra Nevada of California: U.S. Geol. Survey Prof. Paper 73, 222 pp.

Locke, A., Billingsley, P., and Mayo, E. B., 1940, Sierra Nevada tectonic pattern: Geol. Soc. America Bull., vol. 51, pp. 513-540.

Logan, Clarence A., 1934, The Mother Lode gold belt of California: California Div. Mines Bull. 108, 240 pp.

Logan, Clarence A., 1936, Gold mines of Placer County: California Jour. Mines and Geology, vol. 32, pp. 49-96.

Logan, Clarence A., 1947, Limestone in California: California Jour. Mines and Geology, vol. 47, no. 3, p. 263.

Macdonald, G. A., 1941, Geology of the western Sierra Nevada between the Kings and San Joaquin Rivers, California: Univ. California, Dept. Geol. Sci. Bull., vol. 26, no. 2, pp. 215-286.

MacGinitie, H. D., 1941, A middle Eocene flora from the central Sierra Nevada: Carnegie Institute of Washington, Pub. 534, pp. 1-95.

Mayo, E. B., 1937, Sierra Nevada pluton and crustal movement: Jour. Geology, vol. 45, pp. 169-192.

Mayo, E. B., 1941, Deformation in the interval Mt. Lyell-Mt. Whitney, California: Geol. Soc. America Bull., vol. 52, pp. 1001-1084.

Matthes, F. E., 1930, Geologic history of the Yosemite Valley: U.S. Geol. Survey Prof. Paper 160, pp. 31-33.

Moody, C. L., 1917, The breccias of the Mariposa formation in the vicinity of Colfax, California: Univ. California, Dept. Geol. Sci. Bull., vol. 10, pp. 383-420.

Moss, F. A., 1927, The geology of the Mother Lode in the vicinity of Carson Hill, Calaveras County, California: unpublished Ph.D. thesis, Univ. California, Berkeley.

Piper, A. M., Gale, H. S., Thomas, H. E., and Robinson, T. W., 1939: Geology and groundwater hydrology of the Mokelumne area, California: U.S. Geol. Survey Water Supply Paper 780, pp. 32-86.

Potbury, S. S., 1937, The La Porte flora of Plumas County, California: Carnegie Institute of Washington, Pub. 465, Pt. II, pp. 29-81.

Rynearson, Garn A., 1953, Chromite deposits in the northern Sierra Nevada, California: California Div. Mines Bull. 134, Pt. III, Ch. 5, pp. 197-249.

Sander, B., 1950, Einführung in die Gefügekunde der Geologischen Körper: Spruger-Verlag, Wien-Innsbruck, Pt. I, pp. 65-68.

Taliaferro, N. L., 1943, Manganese deposits of the Sierra Nevada; their genesis and metamorphism: California Div. Mines Bull. 125, pp. 277-332.

Taliaferro, N. L., 1951, Geology of the San Francisco Bay Counties: California Div. Mines Bull. 154, [Pt. III,] p. 119.

Tuttle, O. F. and Bowen, N. L., 1950, High temperature albite and contiguous feldspars: Jour. Geology, vol. 58, pp. 572-583.

o

△43460 6 61 3,500   *printed in* CALIFORNIA STATE PRINTING OFFICE